"In an era of populist nationalism, it is important to better understand Indian federalism. This book offers insightful perspectives on federalist cooperation and competition through practical case analyses of important policies such as climate change, the Forest Rights Act, Integrated Child Development Scheme, and more."

John Kincaid, *Professor, Lafayette College, Easton, Pennsylvania, USA*

"This book is a welcome contribution to scholarship exploring the operation of federal principles in the most highly populated federation in the world."

Nicholas Aroney, *Professor Constitutional Law, University of Queensland, Australia*

"This volume indeed lives up to its title in covering the new dimensions. Rekha Saxena has done a wonderful job of bringing together an upcoming crop of scholars to provide insights into current issues of federal practice in India. By looking at intergovernmental relations through the lens of globalization, environmental policy and fiscal federalism the volume provides readers with a better understanding of the imperatives and motivations behind recent policy choices made by governments in India. I highly recommend this volume for anyone interested in contemporary India."

Rupak Chattopadhyay, *President and CEO, Forum of Federations, Canada*

"Timely and well-documented. Young scholars throw light on varied aspects of the multilevel federal system which holds India together."

Balveer Arora, *Chairman, Centre for Multilevel Federalism and Former Professor and Rector-Pro Vice-Chancellor, Jawaharlal Nehru University, India*

NEW DIMENSIONS IN FEDERAL DISCOURSE IN INDIA

This book explores hitherto unaddressed dimensions in federalism studies in India. It traces continuities and changes in Indian federalism since independence, especially economic liberalization. Beginning with the 1990s, due to the emergence of a multi-party system, coalition governments, change in judicial temper and the onset of privatization and globalization in the economy, there has been a trend toward greater federalization in India. However, in the context of a one-party majority in a coalition government since 2014, new aspects have emerged in Indian federalism.

The volume engages with several facets of federalism: administrative federalism; environmental and resource federalism; changing dynamics of fiscal federalism; and multi-level governance. With comparative data and case studies across different states of India, it brings together a range of issues, including Article 356 and its dysfunctions; land acquisition; decentralized governance; tribal rights; the roles of central and state governments; concerns regarding the Citizenship Amendment Act; recent abrogation of Article 370 and 35 A; Delhi and statehood; climate change; MGNREGA; implementation of ICDS; and the cooperative and competitive nature of Indian federalism.

Comprehensive and topical, this book will be useful to scholars and researchers of political science, federalism, comparative federal studies, political studies, comparative politics, public administration, governance and development studies. It will also interest policy makers, bureaucrats, government organizations, NGOs and civil society activists.

Rekha Saxena is Professor at the Department of Political Science, University of Delhi, India.

NEW DIMENSIONS IN FEDERAL DISCOURSE IN INDIA

Edited by Rekha Saxena

Routledge
Taylor & Francis Group
LONDON AND NEW YORK

First published 2021
by Routledge
2 Park Square, Milton Park, Abingdon, Oxon OX14 4RN

and by Routledge
52 Vanderbilt Avenue, New York, NY 10017

Routledge is an imprint of the Taylor & Francis Group, an informa business

British Library Cataloguing-in-Publication Data
A catalogue record for this book is available from the British Library

Library of Congress Cataloging-in-Publication Data
A catalog record for this book has been requested

ISBN: 978-1-138-36150-8 (hbk)
ISBN: 978-1-003-03266-3 (ebk)

Typeset in Sabon
by Apex CoVantage, LLC

DEDICATED TO MY LATE PARENTS . . . FOR FACILITATING THE ART OF DREAMING AND ACHIEVING INSIDE ME

CONTENTS

FIGURES AND TABLES

Figures

Tables

ABBREVIATIONS

AAP	Aam Aadmi Party
AB-NHPM	Ayushman Bharat, the National Health Protection Mission
AIS	All India Services
ART	Article
ASSOCHAM	Associated Chambers of Commerce and Industry
BJP	Bharatiya Janata Party
CAD	Constituent Assembly Debate
CAMPA	Compensatory Afforestation Fund Management and Planning Authority
CEEW	Council on Energy, Environment and Water
CF	Conservator of Forests
CFR	Council on Foreign Relations
CII	Confederation of Indian Industry
CM	Chief Minister
CMs	Chief Ministers
CSR	Centre for Science and Environment
CSS	Centrally Sponsored Schemes
CVO	Chief Vigilance Officers
DDA	Delhi Development Authority
DFO	Divisional Forest Officer
DG	Director General
DGP	Director General of Police
DPCC	Delhi Pradesh Congress Committee
DPSP	Directive Principles of State Policy
FCA	Forest Conservation Act
FDI	Foreign Direct Investment
FICCI	Federation of Indian Chambers of Commerce and Industry
GDP	Gross Domestic Product
GST	Goods and Services Tax
IFS	Indian Forest Service
IG	Inspector General
INGO	International Nongovernmental Organizations

IPS	Indian Police Service
ISC	Inter-State Council
LG	Lieutenant Governor
MCD	Delhi Municipal Corporation
MD	Managing Director
MGNREGA	Mahatma Gandhi National Rural Employment Guarantee Act
MLG	Multilevel governance
MoEFCC	Ministry of Environment, Forest and Climate Change
MOTA	Ministry of Tribal Affairs
NALSAR	National Academy of Legal Studies and Research, Hyderabad
NCRWC	National Commission to Review Working of Constitution
NCTC	National Counter Terrorism Centre
NDA	National Democratic Alliance
NDC	National Development Council
NFSA	National Food Security Act
NGOs	Nongovernmental Organizations
NGT	National Green Tribunal
NIA	National Investigation Agency
NIPFP	National Institute of Public Finance and Policy
NITI	National Institution for Transforming India
NMDC	National Mineral Development Corporation
NRHM	National Rural Health Mission
OTFDs	Other forest dwellers
PCCF	Principal Chief Conservator of Forests
PM	Prime Minister
POSCO	formerly Pohang Iron and Steel Company
PR	Proportional Representation
PSUs	Public Sector Units
PVTGs	Particularly Vulnerable Tribal Groups
RTI	Right to Information Act
SEBI	Securities and Exchange Board of India
SEZs	Special Economic Zones
SRC	State Reorganization Commission
UK	United Kingdom
UPA	United Progressive Alliance
US	United States
USA	United States of America
UT	Union Territory
VAT	Value Added Tax
VOs	Voluntary organizations
WLPA	Wildlife (Protection?) Act
WTO	World Trade Organization

CONTRIBUTORS

Kamal Nayan Choubey is Assistant Professor in the Department of Political Science at Dyal Singh College, University of Delhi, India. He has researched on tribal land rights in India. His published books include *Jugle Ki Haqdari: Rajniti Aur Sangharsh* (2015). He has translated many important books of social sciences into Hindi, including John Rawls's *A Theory of Justice*. He is also a member of the editorial team of the social science journal *Pratiman*, published by the Centre for the Study of Developing Societies.

Sangita Dhal is Assistant Professor (in Senior Grade) and Post-Doctoral Research Awardee at the Department of Political Science, Kalindi College, University of Delhi, India, and has more than 18 years of teaching and research experience. She holds a PhD in political science from the University of Delhi, and her research interests are in public administration, public policy and gender studies. She also has extensive research experience in areas of e-governance and is currently looking at the grassroots democratic process and structural change in rural areas, with a focus on public policy analysis. She has presented research papers in conferences and has contributed articles in several edited books and journals.

Alisha Dhingra is Assistant Professor at the Department of Political Science, Satyawati College (Morning), University of Delhi, India. She pursued her graduate and post-graduate degrees from Hindu College and received an MPhil from the Department of Political Science, University of Delhi. Her research interests include gender studies, constitutionalism, federalism and rule of law. She has presented papers in seminars and workshops. She was part of the Leverhulme International Research Network Project titled "Continuity and Change in Indian Federalism."

Bhav Nath Jha is Assistant Professor in the Department of Political Science at ARSD College, University of Delhi, India. He has contributed articles in several books and reputable journals. Having specialized in Indian government and politics, his current research interests focus on social media and democracy, parliament and the judiciary, e-governance initiatives in rural India and public policy issues.

Niraj Kumar is Senior Assistant Professor, Department of Political Science, Maharaja Agrasen College, University of Delhi, India. With an MA, M Phil and PhD from the University of Delhi, he has been teaching political theory, thought and comparative politics and various other classes for more than two decades. He was UGC Senior Research Scholar in the Department of Political Science, University of Delhi. He has also been associated with some research projects of the University of Delhi. He has half a dozen published articles in books and journals. He is the present editor of *Indian Journal of Social Enquiry*. He has also edited the book *Quality Assurance in Higher Education* (2016).

Amna Mirza is Assistant Professor of Political Studies at SPM College, University of Delhi, India. She is an alumnus of St. Stephen's College (BA program), Hindu College (MA in political science), MPhil and PhD from the University of Delhi and MBA (human resources) from the United Business Institute (Belgium). Her academic initiatives took her to the University of Hongkong, University of Bonn and University of Trinidad and Tobago, among others. She is a recipient of the Godfrey Phillips Golden Ovary Award, Global Indian Award and Bharat Nirmaan Award for Academic learning with Social Entrepreneurship, among others. She has been associated with art and culture festivals and regularly contributes columns to several newspapers and magazines. She has five books to her credit.

Biswajit Mohanty is Associate Professor in the Department of Political Science of Deshbandhu College, University of Delhi, India. His research interests are election studies in India and borderland studies. He has contributed articles in several books and newspapers and well-researched journals such as *Economic and Political Weekly*. He has also contributed an article in international journals on border studies. His latest publication is on ecologic border.

Rahul Mukand is a policy analyst in a top foreign mission in Delhi and Guest Faculty with University of Delhi, India. He received his Master of Philosophy degree in Political Science from Delhi University. After spending over five years as a Research Analyst in premier think tanks, alongside conducting guest sessions in academic institutions in India, he worked as a Political Adviser in the Embassy of Germany (New Delhi). Following this, he received his Master's in Peace and Conflict Transformation from University of Tromso, Norway.

Subhendu Ranjan Raj is Senior Associate Professor in PGDAV College, University of Delhi, India. He is also Honorary Fellow at the Centre for Multilevel Federalism, Institute of Social Sciences, New Delhi. His research specializations include the Indian constitution and politics, federalism, policy studies and the political economy of development. He has published three books – *Coalition Politics in India: Dimensions of Federal Power Sharing* (2009); *The Indian Political System* (2012) and *India in a Globalising Era: A Multidisciplinary Perspective* (ed., 2009), besides contributing more than 30 research articles in leading journals.

Rajesh Ranjan is an independent social researcher and development worker based in India, researching and facilitating people-centric and research-/evidence-based advocacy on rights of indigenous communities, including community rights over natural resources – water, forest, land and minerals and rights and entitlements in pro-poor acts and schemes for the past 15 years, primarily in Chhattisgarh and Jharkhand states of central India. He occasionally contributes articles in mainstream regional newspapers.

Himanshu Roy is Senior Fellow, Atal Bihari Vajpayee Fellowship, Nehru Memorial Museum and Library (NMML), Teen Murti House, New Delhi, India. He was also a fellow at NMML. His publications include *Patel: Political Ideas and Policies* (2018), *State Politics in India* (2017), *Indian Political Thought* (2017 2e), *Indian Political System* (2017 4e), and *Salwa Judum* (2014).

Rekha Saxena is Professor at the Department of Political Science, University of Delhi, India and convener of the Comparative Federalists Research Group. She is also the honorary founder Vice-Chairperson of the Centre for Multilevel Federalism at New Delhi and honorary Senior Advisor to the Forum of Federations, Ottawa, Canada. Her publications include *Varieties of Federal Governance: Major Contemporary Models* (2010), *Mapping Canadian Federalism for India* (2002, edited), *Situating Federalism: Mechanisms of Intergovernmental Relations in Canada and India* (2006, authored), *India at the Polls: Parliamentary Elections in the Federal Phase* (2003), *Indian Politics: Constitutional Design and Institutional Functioning* (2011), *Federalizing India in the Age of Globalization* (2013, co-authored), *Indian Parliament: The Changing Landscape* (2014), and *The Indian Judiciary: The Changing Landscape* (2007, co-edited). She specializes in the study of Indian and comparative politics with special reference to federal political institutions, constitution, center–state relations, federalism, elections and the party system.

Avantika Singh is Assistant Professor at the Department of Political Science, University of Delhi, India. Her research interest is in the study of gender, climate change, decentralization, development and public administration. She has presented her work both in India and abroad. She is the recipient of a full scholarship to undertake her summer school in University of Fribourg, Switzerland. She has widely published many peer-reviewed articles.

Mahendra Prasad Singh is former Professor and Head of the Department of Political Science and former Dean of the Faculty of Social Sciences at the University of Delhi, India. He is National Fellow at the Indian Institute of Advanced Study, Shimla, working on a research project on "Indian Federalism in a Comparative Perspective." He completed his PhD from Alberta in 1975 and has written extensively in the fields of comparative and Indian politics, history of the state in India and philosophy of science and research methods.

Babita Verma is Assistant Professor of Political Science at Laxmi Bai College, University of Delhi, India, since 2006. She has worked extensively on Panchayati Raj institutions and gender, especially in the Uttarakhand region. Previously, she worked with the national commission for women and has published reports on various issues related to gender discrimination. She has also worked with several NGOs working for the empowerment of women in village areas.

FOREWORD

Indian federalism stands at a critical juncture. After a few decades in which multiparty coalition governments were the order of the day (especially between 1996 and 2014), India has witnessed the return of one-party dominance since 2014. The general elections of 2014 were won convincingly by the Bharatiya Janata Party-led National Democratic Alliance. However, unlike the general elections in 1998 and 1999, the BJP now had enough seats on its own and returned to power with an increased parliamentary majority in 2019.

One party dominance is not new to India. In fact, it characterized Indian politics from the first general elections in 1952 until 1989 (leaving aside the period between 1977–1980 and the recurrent losses of power for the Congress in state assembly elections, particularly since 1967). As this volume shows (especially in the introduction by Rekha Saxena), this period laid bare some of the excesses of central power, most notably the recurrent abuse of President's rule for partisan reasons. Coalition politics in combination with the replacement of a state-led by a more market-led economy helped to curb the abuse of central power. States, and perhaps state chief ministers and executives in particular, strengthened their position in the Indian polity, vying for economic investment and for influence in central decision making, even in areas such as foreign policy.

Throughout its three different party systems (Congress one-party dominance between 1952 and 1989, coalition government between 1989 and 2014 and renewed one party dominance since 2014), India has continued to operate on the basis of its more than 70-year-old constitution. Although this founding document was amended more than 100 times, the provisions underpinning its operation as a "union" rather than federal state have remained largely intact. Efforts to put the states on a stronger constitutional footing during the coalition era largely failed. In part this was because strengthening the role of the states was not necessarily on the agenda of either the BJP or Congress, two polity-wide parties with a decisively national outlook and with an essential role in leading the NDA or UPA, respectively. In part this was also because regional or regionalist parties supporting either of these parties were less interested in strengthening the states' constitutional position than in using access to central power as a means to return policy or financial favors to their core constituents. Furthermore, regional parties do not necessarily share the same outlook on issues such as fiscal autonomy. Regionalist parties representing resource-rich states, such as Tamil Nadu, advocate such changes; regionalist parties representing resource-poor states (e.g., the Asom Gana Parishad), may resist more fiscal autonomy.

Although India now finds itself in a one-party dominant system again, the BJP represents an alternative view on India as a "nation" than the erstwhile Congress. Nehruvian democracy was built on the precepts of a secular, state-led or planned and internationally non-aligned polity. It did not necessarily advocate a "federal" but rather "union" state, which, as Ambedkar noted, could be "unitary or federal according to the requirements of time and circumstance." In contrast, the Sangh Parivar, the association of Hindu organizations from which the BJP eventually emerged as its political arm, espoused a different view on the state. This was to be built around the idea of India as a majoritarian Hindu nation. Although this viewpoint presents a direct challenge to the secular moorings of the Indian constitution, it is also not without territorial implications. Early Hindu nationalist writings by Sarvakar or Upadhyadya strongly defended a unitary – not union state – built around 100 *janapadas* (decentralized units of government but under the clear command of the center). They condemned the Indian constitution, built around states and union territories, for its excessive decentralization.

It is fair to say that the BJP's ability to capture power at the level of the states *before* it did so at the center and the party's ability to man some state executives with powerful chief ministers (e.g., Narendra Modi as three-term chief minister of Gujarat) have rendered the abolition of the states and their replacement with 100 *janapadas* unlikely. Furthermore, as the BJP's first prime minister, between 1998–2004, Vajpayee kept the federal spirit alive, in part because he believed in the role of the states, in part because he had no choice as the head of a large central coalition government which relied on the support of regional and regionalist parties.

However, now that the country is headed by its first single-party BJP parliamentary majority, the evidence so far suggests that while India is *not* becoming a unitary state, it also does *not* appear to be moving closer to a "co-operative-competitive federalism." In fact, the constitutionally controversial abrogation of Article 370 in August 2019, on the basis of which Jammu and Kashmir had been accorded a "temporary" special status, is a clear example of non-cooperative central unilateralism. It is justified on the basis of a "state of exception," similar perhaps to what had underpinned the haphazard formation of the NCTC by the Congress a decade prior, shortly after the 2008 Mumbai terrorist attacks (in spite of states' concerns this would undermine their constitutional powers in law and order, as Saxena asserts). It is also in line with Hindu nationalist ideology, which considers a Muslim-majority state as a threat to the idea of India as a Hindu nation. Beyond the case of Kashmir, though, the BJP-led governments since 2014 have upheld the asymmetries embedded in Section VI of the Constitution. These mainly pertain to the tribal populations of the North-East. Yet in attempting to use President's rule to rid itself of politically hostile state governments (as Subhendu Raj asserts) or in using the NITI Aayog as a vehicle to coordinate a strategy among BJP-led state governments to circumvent a Rajya Sabha veto against a proposed federal Land Acquisition Amendment Bill, the central government is not necessarily acting in a cooperative manner, or at best selectively so, that is, only if it can count on the support of states which underline its overall policy agenda and direction. The intention to have at least half of state assembly elections coincide (so far without fruition) provides further evidence of the intent to centralize, as was the shock move in November 2016 to "demonetize" the rupee, which raised serious liquidity issues for the states.

The various chapters in this volume illustrate the mutual ways in which the Indian center and the states are interdependent, in part because the center has more financial and legislative clout but relies heavily on the executive cooperation of the states, especially in areas of concurrent legislative competence. They also lay bare the extent to which globalization and new policy challenges linked to sustainable development, climate change and digitalization (e.g., the more widespread rolling out of Aadhaar) have *increased* the need for intergovernmental cooperation further. Yet although intergovernmental relations and forums for collaboration exist, the position of the states therein as "equal" partners is debatable. The forthcoming recommendations of the XV Finance Commission will provide a further test for the current government's "federal" intent. Will the BJP be minded to uphold the increased financial autonomy of the states which the previous Finance Commission, the XIVth, bestowed upon them? The chapters of this volume give valuable insights into the operation of multilevel governance and the dynamics between the center and the states (and sometimes also local levels of government) in the making and implementation of intergovernmental policy. They illustrate where coordination happens, where it sometimes fails or is inadequately absent and why. They are the fruit of a Leverhulme International Network grant (IN-2013–43) on Continuity and Change in Indian Federalism. The chapters in this volume were first presented during a workshop in January 2017 coordinated by Professor Rekha Saxena of the University of Delhi. This workshop brought together scholars and practitioners from India and the United Kingdom. This book would not have come to fruition without the persistence and organizational skills of Professor Saxena.

As this book goes to press, India is facing many challenges and changes. As outlined here, the age of federalization in India is associated with the emergence of a federal party system and the increased integration of the Indian economy in the global market. However, the gradual demise of the federal party system and its replacement with BJP hegemony also coincides with a partial retreat from globalization. Economically this process is linked to the fallout of the global financial crisis in 2008–2009, which also slowed the growth path of the Indian economy. Politically it is tied to the rise of national protectionism and with it the advent of populist nationalism and the *reassertion* of the state in relation to the market. As I write this foreword, the world is in the midst of a pandemic. Fighting COVID-19 will require cooperation between the Indian center and the states, given that health is a concurrent competence. Emergencies such as these (as is redressing the economic fallout from putting a country into a prolonged national lockdown) not only increases the size and role of the state relative to the market; in federal states, events such as these also often increase the size and role of the center relative to subnational (state) governments. Some of these changes may be difficult to unwind once the emergency subsides and life assumes a degree of normalcy. This makes studying center–state relations in India today a highly interesting but also rapidly evolving subject.

Wilfried Swenden
Professor of Politics, University of Edinburgh, Scotland, UK
and Leverhulme International Network Coordinator of
Continuity and Change in Indian Federalism (2014–17)

PREFACE

This book explores the new dimensions in Indian federalism. It also traces continuities and changes in the nature of federalism in India since independence. With the advancement of coalition governments and the rise of regional parties, many scholars started talking about the strengthening of federalism in India. Indian federalism has been reshaped by the outcome of the 2014 general elections, in which the BJP led by Modi has emerged as a majority party after three decades (during which no party was able to get a majority in the Lok Sabha). The BJP also formed government in most of the states of the Indian Union, either as a majority party government or as the head of coalition governments. This reminds us of the one-party dominant system under the tutelage of the Indian National Congress from the 1950s to 1980s. Despite the rise of BJP as an unmatchable force, a set of regional parties in some states of the South and East have continued to hold the ground, and in some recent assembly elections in Punjab, Madhya Pradesh, Chhattisgarh and Rajasthan, the Congress party was able to form the government. The BJP has been cautious to keep the NDA coalition intact amidst the challenge of a number of regional parties, and the Indian National Congress has largely remained disunited in a coalitional front.

The replacement of the Planning Commission with the National Institution for Transforming India (NITI) Aayog in 2015 has had substantial implications for shared rule and the nature of cooperative federalism in India. The cabinet resolution of 1 January 2015, which led to the birth of NITI Aayog, declared the "shared vision of national development priorities, sectors and strategies with the active involvement of the states in the spirit of 'co-operative' federalism" as the central objective of NITI Aayog.

Another significant change in structure of tax reforms is the shift from the Value Added Tax (VAT) to the Goods and Services Tax (GST), which seeks to make India a common national market, with the GST Council comprising the union finance minister as the chair and state finance ministers as members, with two-thirds vote for the states block, as enshrining the spirit of cooperative federalism.

In the 2019 general elections, the Bharatiya Janata Party (BJP) again formed a government securing 303 Lok Sabha seats, breaking its own electoral record of the 2014 election. The Indian National Congress remained on the margins of electoral performance, with just 52 seats. What is significant to note is the fact that in the 2019 elections, the Congress party was completely wiped out from 18 states, and the vote share of BJP has risen to 50 percent in 13 states and union territories. This has led the Bharatiya Janata Party to emerge stronger than ever before, which is visible in many of its actions after its electoral victory in 2019. Most significantly, the recent enactment of the Jammu and Kashmir Reorganisation Act 2019, an abrogation of Article 370 and 35 A of the Indian Constitution (an asymmetrical provision), has significantly reordered the federal structure in India. The Ram Janma Bhoomi movement, spearheaded by the BJP, has also culminated into the beginning of the construction of the Ram temple in Ayodhya. However, the federal balance of forces is still sustained, as in some states, regional parties have held their ground and stymied the extension of the BJP in parts of southern, eastern and western India.

The political leadership at the state level has opposed Citizenship Amendment Act 2019. This is interesting in light of the fact that India does not have dual citizenship, and the Constitution entrusts the Parliament to make laws on citizenship, giving no power to state legislatures in this regard. The conflict that the CAA has generated might become a prototype for future contestations over the federal question in India.

In light of the spread of a pandemic like coronavirus, the spirit of cooperative federalism becomes important to deal with such situations. Federalism in India has been both competitive and collaborative, and the need of the hour is to strengthen cooperative intergovernmental relations.

ACKNOWLEDGMENTS

This book has been a long six years in the making. It is an outcome of a three-year project titled Continuity and Change in Indian Federalism (2014–2017), funded by the Leverhulme Trust in UK. The project was an institutional collaboration between six universities, three from the United Kingdom, namely Edinburgh, Bristol and Nottingham, and three from India, namely the University of Delhi, Burdwan University and Hyderabad Central University. The principal applicant of the project was Wilfried Swenden from the University of Edinburgh. The chapters in the book were part of the subtheme on intergovernmental relations in India and were first presented at a workshop organized by the Department of Political Science, University of Delhi, India.

First of all, therefore, I would like to thank the Leverhulme Trust for the financial support that enabled us to do fieldwork and research. Next I owe my thanks and gratitude to Wilfried Swenden for associating me in the project as a lead coordinator from University of Delhi on the subtheme "intergovernmental relations" and for his understanding and support during the course of the project. I would also like to thank all other subtheme coordinators, especially Harihar Bhattacharya, K. C. Suri, Katherine Adeney and Andrew Wyatt for their advice and guidance. I owe special thanks to Rajesh Jha, Chandrachur Singh, K. K. Kailash, K. K. Suan Hausing, Chanchal Sharma, Abhay Prasad Singh, Tanvir Aeijaz, Sushant Kr. Kanungo, Avantika Singh, Ashutosh Kr. Jha, Jiya Lal, Amna Mirza, Bhav Nath Jha, Annapurna Sharma, Prakash Kr. Patel, Jiya Lal, Sai Srinivas, Bhavtosh Bhaskar, Anant Prakash, Utsav Kr. Singh, Vasavi and Mohan Kumar for their contributions to the project in various ways. I would also like to acknowledge Harsh Meena, Alisha Dhingra and Rounak Kumar Pathak in particular for their research and technical assistance.

Thanks are also due to Ujjwal Kumar Singh and Navnita Chadha Behera, both former heads at the Department of Political Science at the University of Delhi for their encouragement and support during the course of the project. I would like to thank my other colleagues in the department who were also associated with the project, particularly Ashok Acharya, Sangit Ragi and Rajesh Dev. I owe an enormous debt to M.P. Singh, who nurtured my interest in federal studies and encouraged and guided me through all these years. Special thanks to V. K. Ohri for being a standard reference for legal queries and for his continuous guidance and support.

My thanks are also due to Balveer Arora, Rupak Chattopadhyay, Kamal Nayan Kabra, C. P. Bhambhri, Imtiaz Ahmad George Mathew, Rajeev Dhavan, Prashant Bhushan, Ash Narain Roy, Douglas Verney, Bidyut Chakrabarty, Subrata Mitra, E. Sridharan, John Kincaid, Nico Steytler, Francesco Palermo, John Mcgarry, Nicholas Aroney, Cheryl Saunders, Will Kymlicka, Ian Greene, Kathy Brock, Jonathan Rose, Paritosh Kumar, Louise Tillin, Leena Rikkila, Surya Dhungel, Radu Cârciumaru, Bimalendra Nidhi, Chandan Mitra, D. Raja, Manish Tewari, Ashwini Kumar, Ashutosh Kumar, Gull Wani, Ajay Kr. Singh, Kumar Suresh, Pampa Mukherjee, Sushma Yadav, Rumki Basu, Sandeep Shastri and Anil Singh for their interest in my work and their critical intellectual feedback.

I would also like to acknowledge the late Mandakini and M.M. Sankhdher for their help and support during my arduous times. Acknowledgement is also due to late Akhtar Majeed, B. D. Dua, Ronald Watts, Hugh Thorburn and C.E.S. Franks for their help and support. I collectively thank all my friends, colleagues, research scholars and students, who have been a great support for me; I have immensely benefitted from my discussions with them, but it is difficult to single them out individually. In addition, I would like to thank Ada Munns, S. K. Bagga and Mamta Shah for their secretarial assistance.

I am grateful to Shashank Sinha, Antara Ray Chaudhury and Rimina Mohapatra at Routledge for patiently bearing with me in getting this book published.

Finally, I owe all my family members a huge debt of gratitude for their all time valuable support and affection, especially my sister Renu and brother-in-law Shailesh; brother Deepak and sister-in-law Anu; nephews and niece Abhishek, Shashank, Rishabh, Vansh and Sukriti. I am grateful to my brother Ajay, whose memories have always extended moral support to me. I am also indebted to my cousins Kavita, Mamta, Amita and kido Rhea for their care and support. Further, I will always be obligated to my late Aunt Uma and Uncle R.B. Saxena for being pillars in my journey, and my Uncle Shyam Babu Saxena for his encouragement. In the end, my acknowledgment stands unfinished without paying panegyrical tribute to the fond memories of my late parents, Ram Babu Saxena and Asha Saxena, as they are the source of my vigor in all my achievements till date.

Rekha Saxena

INTRODUCTION

Dialectics of collaborative and competitive federalism

Rekha Saxena

India has a plethora of religions and languages, but it is the mystifyingly diverse cultures that sway the everyday lives of India's diverse peoples. Accordingly, varied campaigns' political aspirations and constitutional demands emerge out of social and political interests kindled by those diversities. Thanks to the flexible nature of the Indian Constitution, these varied aspirations and demands find safe haven through politics and through the use of constitutionally provisioned amendment power.

However, accommodation of a wide range of demands coming from diverse group of people by Indian federalism within its constitutional structure was an arduous task (Saxena 2012a). The roots of Indian federal structure can be drawn back to the Government of India Act 1935, which was motivated by a British Raj system that was revenue based. Post-independence, moderate features of the 1935 Act were incorporated in the new Constitution of 1950, which ultimately helped in creating a model based on planning, taking inspiration from America's New Deal and Soviet planning. Come 1967, we witnessed the transition from planned development model to a "cooperation-bargaining" model as a consequence of the 1967 general elections in which, for the first time, most of the states were governed by political parties different from the ruling party at the center (Dhavan and Saxena 2004). Looking at the Indian case in historicity, we find that there was a pertinent presence of system of intergovernmental fiscal arrangements in India over a period of almost 50 years. From the equal allocation of services to instituting a workable system capable of resolving issues between the center and states, this arrangement has achieved them all. In this way it has resulted in a cohesive system in a large multicultural country like India. The planned development model was displayed ostentatiously during the period of emergency in order to conceal its autocracy. But since the commencement of unstable coalition-based government at the center in 1977, and particularly after 1989, India has understood the need to redesign and renovate its federal design, bringing in the element of bargaining. Such tendencies have only made federalism both frail and sturdy.

In the era of liberalization, an open Indian economy has brought about significant changes in the fiscal federal arrangements. Such changes constituted certain challenges, which included the introduction of taxes against public enterprise revenues, overcoming loss from customs, and evolving a destination-based value-based tax in the country (Rao 2006). The neoteric globalized economy has transformed the nature of states to an efficient performance of its role with respect to social service delivery and creating physical infrastructure. Furthermore, in order to prevent the stalling situation of fiscal deficit and higher debts, fiscal discipline at the subnational level has to be ensured by the fiscal system today. Another noteworthy situation has consequently emerged with the advent of coalition government at the center and states, which compels the political leaders to adopt populist policies without paying heed to the fiscal consequences they might have. The presence of regional parties as the major constituents of the coalition at the center has led to several asymmetric arrangements, which in turn adversely affect the fiscal arrangements.

India's federal constitution

Constitutions all over the world are passing through a dual phase of both "being a constitution" and "becoming a constitution" (Dhavan and Saxena 2004). The functional aspects of constitutions all over the world have seen resistance from both time and factors that had prevailed and factors that are new. Under such pressurizing tendencies, the emergent constitutions take over their predecessors. Akin to various constitutions all over the world, there are varied and complex provisions with respect to bringing out an amendment in the Constitution of India. Such provisions alter different facets of the Indian federal system, and it requires the bill to be approved by a two-thirds majority in each house of the Parliament followed by ratification by one-half of the legislatures of the states. In the Keshavnanda Bharti case (1973), the apex court of India threw a restraining notice on the Parliament of India to not control its amending power with respect to the "Basic Structure" of the Indian Constitution, which is inviolable. Without a shadow of a doubt, the federal feature of the Indian Constitution forms a key ingredient of the Basic Structure Doctrine. However, most of the amendments that have been made to constitution till date have affected the federal system in India. Furthermore, it is noteworthy to mention here that such constitutional altercations are the product of varied pressures and practices and the exercise of executive and legislative power, in addition to the many amendments.

In the past six decades, we have witnessed drastic changes in essence and functional aspects of federalism in India, despite the changes brought about by prevailing situations. The Indian federation has witnessed two concurrent changes in the 1990s: transformation of the one-party system to a multiparty system and transformation of a mixed economy established at independence to a market-oriented economy (Singh and Saxena 2011; Pai 2013). Keeping in mind the geographical requisites of India, the blueprint of the constitution has been recreated. The model has experienced change in its entirety. This can be illustrated from the kind of changes that have happened with the states. Many have split into smaller new ones, endorsing the claim that even states are ephemeral, while in some states, especially in tribal areas, governance apparatuses of intermediary configurations have been set up. The introduction of the three-tier structure of governance in 1992, with the passing of 73rd and 74th amendment of Indian Constitution, completely overhauled the federal model. It constitutionalized the third tier of federal structure and consolidated multilevel federalism. Peter DeSouza has argued that this resulted in a new phase in Indian democracy that he terms as the "second wind of democracy" (DeSouza 2003) in which the federal model completely underwent umpteen changes, both theoretically and operationally, effectuated by uses and misuses of the power by the executive and legislative branches. Indian federalism has been dynamic and rewritten several times through constitutional amendments by the legislature and constitutional interpretation by the judiciary. The abrogation of Article 370 has provided the opportunity to rewrite Indian federalism. I think it was the failure of successive central and state governments to maintain a balance between self rule and shared rule that culminated in the abrogation of this asymmetrical provision. Further, Tillin (2016) argues that the Supreme Court had effectively permitted the weakening of Article 370 over time by legitimizing revisions to the Article that extended much of the Constitution to the State. The erstwhile state of Jammu and Kashmir has been divided into two union territories, Jammu and Kashmir (with a legislature) and Ladakh (without a legislature). Union territories (centrally administered territories) have been created in India from time to time for varied reasons. The reason behind the creation of Jammu and Kashmir and Ladakh has been the longstanding security problem and lack of development. It remains to be seen whether these problems can be resolved through the union territory model or an alternative arrangement is required.

It is the process of decentralization within an exceedingly centralized political system that is responsible for the creation of the Indian Union. This is in contrast to the United States where erstwhile autonomous states came together to form a union. This dissimilarity makes it easy to comprehend as to why there are more centralizing tendencies in India than in America. Another reason that could be cited for the high degree of centralizing tendencies within Indian federation is four decades of one-party dominance, both at the center and the state. However, beginning in the decade of the 1990s was a major turning point in the nature of the party system, and the political climate since then has undergone a major churning, thanks to the emergence of large number of political parties. This ultimately resulted in a multi-colored party configuration in the state party system. The neoliberal economic reforms that were introduced in 1991 brought significant transformation in the political economy of federalism in India. Consequently, governments in India felt compelled to bring about the accommodation of market forces and civil society institutions in the process of planning and economic management. Furthermore, today the union government cannot afford to ignore states on the questions of issues that have center–state relevance. Also, with respect to the formulation and implementation of Centrally Sponsored Schemes, the union government has to consult the state governments. This is the consequence of federal coalitional governance.

Continuity and change in Indian federalism

In the context of center–state relations, attempts have been made by contemporary scholars to chart the fundamental nature of changes in India's policy regime. According to Lawrence Saez, there was quick transition in federal relations from intergovernmental cooperation to inter-jurisdictional conflicts among the states with the commencement of economic liberalization policies in 1990s (Saez 2002).

The 1991 liberalization not only brought about a shift from public to private sector and from state to the markets, but it also reshaped the part of regional states vis-a-vis the central government. The process of federalization has enabled the political sustainability of India's reform process by relocating the opposition to the state level, as argued by Rob Jenkins (Jenkins 1995)

In this context, M.P. Singh argues that there is a mutual reinforcement between federalism and neo-economic reforms, and both together augment the autonomy of state governments and the domestic/global private sector. The nature of the Indian economy created a new phenomenon of a regulatory state and sectoral federalism more horizontal than vertical, mainly due to the replacement of direct administrative control of the economy by autonomous regulatory bodies in the new political order (Dua and Singh 2003).

Regional states adopted diverse ways of shaping and implementing economic reforms, and this has been analyzed by scholars. Also noteworthy is the change and continuity visible in the center–state relations in the post-1991 era. The nature of competition in the post-reform era was characterized by cooperation with other states and domination by the center and was different from what it is at present. The nature of competition in the post-reform phase has shifted from vertical (in which states contended with each other for centrally determined resources) to horizontal (in which states contest with each other more for resources from varied actors) (Sinha 2004). This symmetrical horizontal competition has unleashed "processes of diffusion and learning by copying among a greater number of states" than before. Though the number of competing states has increased and there is ostensible junction of policies and discourses with the increasing number of competing states, there are regional disparities in outcomes (ibid).

In a perceptive analysis, the Rudolphs argue that the 1990s economic reforms in India have brought about a transformation in the nature of Indian economy. It has shifted from a command economy to a federal market economy that facilitates a wider "sharing of sovereignty" within the state and the market. In the new era, states have started to command a larger share of economic sovereignty, unlike what it was during the centrally planned economy. The barometer for judging their economic performance depends more upon what they do for themselves. Today, the vital place for private investment is states. There are mutual advantages and benefits that the states can derive from their cooperation over a longer period, and henceforth they should forgo of the short-term benefits. Most importantly there is a transformation of the union government from being an interventionist and guardian of a "federal market economy" to one enforcing "fiscal discipline" and ensuring transparency and accountability (Rudolph and Rudolph 2001). In the era of globalization, the states in the Indian federation have emerged as forceful entities to be reckoned with. It needs to be emphasized that some states are governed by political parties ideologically incongruent to globalization and macro-reforms (Bhattacharya and Konig 2016).

Since 1994, meagre changes have emerged in the formal architecture of Indian federalism. For example, there has been no modification in the list of exclusive central powers since 1994 (the 88th amendment made the center responsible for taxes on services) with respect to concurrent and state lists and no amendments were approved; however, alterations to the distribution of tax resources between the center and the states (underpinned by constitutional amendment) have taken place, as specified in Part XII of the constitution (Swenden 2014).

Nonetheless, in the exercise of the powers conferred to the center and the states, there have been significant changes. These have been referred by Singh and Saxena (2013) as indexes of federalization, specifically (1) a sharp descent in the imposition of President's rule (made possible in part through Bommai Supreme Court rulings which especially sought to prevent its abuse for party political reasons by restricting the use of PR); (2) a descension in the occurrence of central intervention in the legislative process at the state level, explicitly visible through governors' exercise of power to reserve a provincial bill for consideration of the president; (3) states proclaiming their rights in the conduct of foreign policy, if it impinges upon state interests (the WTO treaty in 1995 affecting agriculture, a state competence; successful lobbying by TMC-ruled West Bengal to prevent the Indian-Bangladeshi water-sharing accord) – similarly, chief ministers (CMs) and finance ministers of the state governments have been more often incorporated in foreign missions to attract capital investment; and (4) the instances where presidents and governors have more constantly used their discretionary powers in relation to state governments and legislatures (where these are not subject to explicit parliamentary consent) are quite contradictory and call for considerable investigation.

The pertinent question is how one can bring about informal change in federal dynamics, seeing the minimalistic level of formal change. Going by these suggestions, one possible way is to bring about change in both practice and frequency with which constitutionally allocated powers are used (e.g. President's rule). Conceptually, Benz and Broschek's (cited in Swenden vv) interpretation of change in federal regimes as occurring in three different arenas, can also be adopted. These arenas are (a) in the vertical differentiation of authority between territorial levels of government (in our case between central and state levels – authority is only partially linked to formal powers; it is of course also linked to informal power and the (non-)use of formal powers; (b) the horizontal structure between constituent units (as influenced by horizontal differences in fiscal, political, infrastructural or cultural capacity), as well as formal, that is, constitutional asymmetries; (c) the intragovernmental arena: the structure of governments constituting a federal system, organizational features of parties and party systems or democratic and nondemocratic forms of government.

Cesar Colino (cited in Swenden 2014) has conceptualized about formal and informal change. According to him, the formal federal framework consists of three dimensions: constitutional design (which determines the degree of intrastateness, i.e., the kind of regional involvement in federal decisions, the kind of power distribution – "functional versus sectoral," the "concurrency or exclusivity of legislative competencies"), the existence of asymmetry in statutory capabilities and the assignment of residual powers. Outside the constitution but also belonging to the formal framework, we find intergovernmental structures and resources that corroborate with the degree of interdependence. This variable reflects the "(de)centralization of spending and resources, the revenue autonomy and fiscal responsibility, the extent of administrative decentralization, the existence of vertical or horizontal intergovernmental structures for decision-making and the nature and formalization of cooperation bodies" (Colino, 2013: 55 cited in Swenden 2014). Intergovernmental decision-making rules determine the degree of hierarchy in intergovernmental relations, and these can be measured by considering the "type of existing rules of initiative in intergovernmental bodies, their binding or voluntary character of joint-decisions, the aggregation rules for decision-making, the rules on conflict resolution and the general ambiguity of rules" (Colino 2013: 55 cited in Swenden 2014). One is able to distinguish between first order and second order formal change by reorganizing formal change along these three layers. In the Indian context, the advent of liberalization put together with emergence of coalition forces has slanted the equilibrium on the way to a more "decentralized federalism" in which the states have a new concern for their distinctive growth. Though power was still located in the central government, a new environment was created in which states would compete with each other and with central government for their voice in policy issues of their concern (Swenden 2014).

The constraints emanating from the constitutional design are compensated for by the informal structures. The very essence of federal relations is shaped through collaboration and competition in the form of interaction and joint decision-making styles, solidarity-oriented strategies of governmental actors and party political and territoriality claims represented through intergovernmental coalitions. Colino's framework goes beyond the formal – especially constitutional – parameters and proves beneficial in paying due attention to formal rules at the meso- and microlevel of center–state interactions as well as their informal dynamics. It is in the separation of formal from informal and assessment of interrelationship between both, where the real struggle lies. In some sense, the change in informal relations relates to the rise in the degree of interstate competition that has followed from the party-incongruent composition of state governments in relation to each other and in relation to the center and the liberalization of the Indian economy resulting into the gradual retrenchment of the center from direct intervention. The instances where liberalization replaces socialism or monetarism replaces centralized planning results in new ideas, which in turn produces different codes of intergovernmental conduct or behavior (ibid).

Going by the current controversies surrounding the subjects of governance, economic and foreign policies, a novel dimension has emerged in the paradigm of center–state relations. The new issues – ranging from water law to land rights, climate change, energy policy, conflict between development and environment, urban governance, fiscal management, fiscal equalization, metropolitan governance and designing an immigration and integration policy – are what form part of the governance deficit.

With the expanding participation of nonstate actors in the arena of governance, many more changes have emerged. Furthermore, steering, coordination and networking have become the new facets of state functions, evolving from what earlier used to be command and control, as a result of which ensuring democratic accountability has become quite a bit of a hassle in the varied types of structures of governance. By imparting constitutional status to the Panchayati raj institutions through 73rd and 74th amendments, a new wave of interest has developed in multilevel governance. Thereupon, there has been a surge in the prominence of varied intergovernmental institutions, viz, the Securities and Exchange Board of India (SEBI) or National Green Tribunal (NGT). Today, a cooperative relation between the center and states has been an essential prerequisite of policy concerns.

Discord between the center and states

On issues such as the Lokpal and Lokayukta Bill, Naxalism, national security and the NCTC, foreign direct investment (FDI) in retail trade, land acquisition, food security, corruption in Centrally Sponsored Schemes like MNREGA, NRHM, the Indo-Bangladesh agreement on the distribution of Teesta river waters, GST, NITI Aayog, etc., there has been a lot of discord between the center and the states. There are three lists – viz the union list, state list and concurrent list – in the seventh schedule of the Constitution, which talks about the federal distribution of the responsibilities, in which the aforementioned matters have acquired an indispensable significance. However, the formal division of powers are rendered intelligible because of the intermeshing of these three lists into the constitutional provisions and functional relations. A line of demarcation between exclusive jurisdictions of two levels of government that could have been drawn has become impossible. This can be illustrated from the fact that when the central government signed the WTO treaty, it led to the filing of suits by at least four state governments before the Supreme Court, as according to the latter, this treaty compromised their exclusive jurisdiction in agriculture.

The domestic practices of unilateral executive power inherent in the Article 253 have invited resentment from the states, making one wonder whether such practices are at odds with fluctuating internal and external dynamics. Federal foreign affairs, which had been an exclusive domain of the central government, is now opened to the states due to geography, coalition politics and economic linkages. The changing contextual reality together with the Indian federal practices actually determines the theory of center–state relations defined in the constitution (Saxena 2012b).

The contestation is over the unilateral exercise of power by the executive in signing international treaties. The vagueness of the Indian scheme of center–state relations does not allow for a dynamic federal polity based on center–state relations. This matter and the power emerging out of it, which lies in the domain of Parliament, should be exercised while keeping in mind the interest of the states and their respective citizens. In one Supreme Court judgments, Justice Ganguly came to an opinion that courts can exercise the power of judicial review with respect to unconstitutional and unquestionable executive powers. Its imperative on the part of executive to parliamentarian and federalize this treaty-making power, thereby endorsing transparency and accountability in its exercise. This was also seconded by the left parties in their opposition of FDI in retail sector (Saxena 2007).

The Teesta water sharing agreement was nearly settled during UPA II regime when the deal was being signed between India and Bangladesh in September 2011 on an equal sharing basis. Nevertheless, the deal was deferred when the CM of West Bengal, Mamata Banerjee, declined to escort PM Manmohan Singh to Dhaka, arguing that the welfare of Bengalis had been ignored and the draft agreement was substantially distorted from the one that was presented before her, although the PM contended that Mamata Banerjee was properly briefed by the National Security Advisor Shiv Shankar Menon almost a month before the planned visit to Dhaka. The hasty act of Mamata Banerjee not only embarrassed the PM but also disrupted the process of bilateral agreements between the two countries. This was an impediment to what scholars describe as constituent or para-diplomacy. The transformation of the party system from one-party dominance to multiparty configuration and initiation of neoliberal reforms have allowed states to expand their role in the domain of foreign policy. Therefore, there is a need for active consultations between the union and the states on issues relating to foreign policy (Saxena 2007; Singh and Saxena 2013; Basu 2016).

Similarly, on the issue of security, a controversy emerged over the creation of NCTC. The chapters in this book do not discuss the security dimension of the discourse on federalism in India. Therefore, I will lay out the controversy over NCTC in detail to map out the federal dimension of the security discourse in India. Multiple chief ministers' conferences convened by the NDA and UPA-I governments were used as a platform to reach a consensus between the union and the states on the establishment of a federal agency with sufficient powers to investigate interstate crimes. Nevertheless, these conferences proved fruitless, as the chief ministers did not achieve consensus with respect to defense of the states' rights in relation to law and order (a state subject) under the Constitution. Toward the end of its term, the UPA- I government demonstrated a nimble move by getting National Instigative Agency (NIA) Act passed by a voice vote in Parliament, especially in the context of the 26/11 Mumbai terror attacks. There was no opposition whatsoever within the Parliament, and the chief ministers did not opposed either, probably in view of the traumatic terror attack.

The neoteric decisions of the UPA-II government to set up the NCTC by an executive order and arming it with the power to carry out searches and seizures under Section 43 A of the UAPA has already invited and ignited a lot of controversies. Such an offensive move caused a great deal of discontent among the state governments. They could see their jurisdiction with respect to policing powers of law and order granted to them under the Constitution of India as being compromised. Further source of agitation arose from the nonconsultative unilateral actions taken by the center without taking their opinion into consideration. The chief ministers who opposed the UPA government's decision generally included non-UPA-ruled states like Biju Janata Dal in Odisha, Janata Dal (United) in Bihar, AIADMK in Tamil Nadu and BJP in Gujarat, as well as Trinamool Congress in West Bengal, which in fact is was an ally of the ruling UPA-I in Delhi. Their concerns were particularly directed toward the likelihood of abuse of the power of arrest and search granted to the NCTC division of the Intelligence Bureau against political opponents by labelling them as terrorists.

Nonetheless, it is pertinent to mention here that the plea of union Home Minister Chidambaram for an expeditious action against interstate terrorism was valid. Giving "minimum powers" to NCTC to counterterrorist operations and prevent the terrorists' plans were the line of arguments made by him. The powers granted under Section 43(A) must be read in conjunction with the obligation under Section 43(B) of the act to hand over the suspect after arrest to the nearest police station under the state government.

Worth mentioning is the fact that the Indian version of NCTC is stronger than the US version, as in the United States, the National Counterterrorism Center works under the Director of National Intelligence and does not have the power to direct the execution of counterterrorism operations. Furthermore, it also lacks powers of arrest, interrogation, investigation and prosecution. In India, Entry 2 A in List I of the Seventh Schedule, inserted through the 42nd Amendment Act of 1976, provided that in extreme cases, the center can deploy military or paramilitary forces in aid of civil order in a state. Moreover, according to Article 355, the center is charged with defending states from external aggression and internal disturbances. It is clear from the aforementioned points that the union government was no-consultative and unilateral in setting up NCTC, which should have been avoided (Saxena 2012c). The political atmosphere of the country has changed, with the emergence of regionalization and federalization indicating the need for union–state consultation. It was the recommendation of Parliament's Standing Committee for Home to the union government to not to go ahead with the creation of NCTC unless consultations were sought with the states. Also, NCTC's operational apparatus would include anti-terrorist squads of the states and all state DGPs, as agreed by the Home ministry. However, the deadlock continued yet again at the chief ministers' conference held on 5 May 2012, when the center failed to convince the chief ministers. The states were of the view that a subcommittee of the chief ministers headed by Jayalalitha should be set up so that it could review the issue and submit its report.

Convening a meeting of Inter-State Council (a constitutional body) is the best way to end this stalemate, as the decisions taken at such a meeting would have both constitutional and political legitimacy. Setting up a standing committee of the ISC made up of the home minister and chief ministers on rotation and providing them with authority to deal with crisis situation, is another solid solution. The Punchhi Commission in its report endorsed appropriate amendments in order to make the ISC a "credible, powerful and fair mechanism for management of inter-state and centre-state differences" (Punchhi Commission Report 2010).

Another illustration of federal controversy emerged with the introduction of the 116th constitution amendment bill on 22 December 2011 to facilitate the creation of Lokpal. The issues that were staunchly opposed were a reservation for minorities in Lokpal, control of the Central Bureau of Investigation by the union executive and obligatory provisions for states to establish Lokayuktas. The bill pursued granting constitutional status to Lokpal and Lokayuktas. This alleged move of the center to encroach upon states' domain was resisted by both the opposition and states.

Since there is barely a party that does not have corrupt and criminal legislators, the parties exhibited a determined opposition to the bill. Moreover, in addition to the cold reception to a strong anti-corruption law, there was a supposed attack on the federal structure of the constitution. BJP and all regional parties opposed the bill, as it was encroaching upon the jurisdiction of states. The creation of Lokayukta also was a concern with respect to the civil services of the states. Under Article 252 of the constitution, on a request from two or more states, the union government can legislate exclusively for them. This would have enabled state governments to bring about their own versions of the Lokayukta Acts freely, largely within the proposed outline (Saxena 2012d).

In order to effectuate its commitment to sign and ratify the UN Convention on Corruption, the government decided to bring a bill under article 253 to create a binding effect on all the states. What is more astonishing is that no attempt was made to create a union–state consensus on this issue despite knowing the sensitivities of the states in this phase of greater federalization (in the sense of regionalization) of the Indian polity.

As per the arguments given by the government, Article 252, if invoked, is nonapplicable to the current case because the subject of the proposed legislation must be in state list of the Seventh Schedule of the Constitution. And that corruption is a criminal act; therefore, they have power to legislate under entry 1, 2 and 11- A of the concurrent list. Such arguments have a non-persuasive nature because proposed legislation is more applicable to civil servants and political class than to the criminal code, per se.

Moreover, placing Lokayuktas under a mandatory central law as proposed in the new Lokpal bill would simply take away role of the state in the appointment of a Lokayukta and his staff. According to one clause of the bill, funds for state Lokayukta would be given from the consolidated fund of the state that is not subject to annual budgetary authorization by the state legislature. This reflects the controversial nature of the bill.

Further controversies had been evident in the states of Gujarat and Karnataka, which were ruled by opposition parties. In Gujarat, the state government had challenged the appointment of R.A. Mehta as the Lokayaukta by the governor in the Gujarat high court. The court upheld the appointment of the lokayukta, but the government has indicated that they will go to the Supreme Court in appeal against the high court ruling. Likewise, in Karnataka the confrontation also continued between the chief minister and the governor over the appointment of S.R. Bannurmath as the lokayukta. Thereupon, the UPA government finally capitulated to the opposition Bharatiya Janata Party's demand to refrain from going ahead with its Lokpal Bill under Article 253 of the constitution (Saxena 2012d).

The most recent conflict between the central and state governments has occurred in the context of the Citizenship Amendment Act 2019. Several state governments occupied by opposition parties or partners of the Bharatiya Janata Party (BJP) have declared that they would not implement the law. Moreover, in a rather unparalleled move, the Legislative Assembly of Kerala went to the extent of passing a resolution stating that the law "contradicts the basic values and principles of the Constitution." Certainly, the resolution is only emblematic and has no legal implications. Article 256 of the Constitution requires the state government to ensure implementation of the laws made by Parliament. If the state government fails to do so, the central government is authorized to give "such directions to a State as may appear . . . to be necessary." Refusal to implement the law even after the center issues directions would authorize the president to impose President's rule in those states under Articles 356 and 365. Another controversy emerged in West Bengal, where the state government has put up anti-CAA advertisements on its websites and the Calcutta High Court directed the state government to remove those advertisements. Though neither the refusal to implement Citizenship Amendment Act nor the official protests registered by state governments carry legal force, it is interesting to note that such proactiveness on the part of some state governments is the consequence of the emergence of a dominant party at the center.

Outline of the book

The first chapter (Subhendu Ranjan Raj) talks about Article 356 and its misuse in the federal system. The instrumentality of Article 356 and its accompanying provisions have been responsible for collapsing popular democratic governments, creating a discontinuity in governance, and stirring up public outcry against it. Inasmuch as Article 356 was inspired in the earlier years by Section 93 of the Government of India Act, 1935, which provisioned for a "controlled democracy" the need of the center to intervene in the legitimate process of governance, and government formation in the states has only increased of late with the changes in the Indian federal construct brought about by a neoliberal/globalized economy, the upsurge of dalit identity and the assertion of regionalism with the growth of lower caste parties in recent years. Rather than any aberration in the constitutional provision, it is manifestly clear that, as against the intentions of the Constitution-makers, the majority of problems associated with the ignominious Article 356 have been created by the vested interests and political practice of successive governments at the center. Except for a few who rose above partisan considerations, most of the governors have misused their constitutional role. The chapter discusses the recent happenings in Arunachal Pradesh and Uttarakhand, which are the most recent amplification of such misuse. In consonance with the idea that Article 356 is crucial for the misuse of the federal architecture in the constitution and that the center is prone to do so in its current form, the chapter argues for reforms in the role of governor and in the provisions giving explicit dominance to the center's will and intent that urgently need to be watered down in the light of federal accommodation.

The Scheduled Tribes and Other Traditional Forest Dwellers (Recognition of Forest Rights) Act, 2006, popularly known as the Forest Rights Act (FRA), was enacted in 2007 by the Ministry of Tribal Affairs to correct the "historic injustice done to forest dwelling communities." The second chapter (Rekha Saxena and Rajesh Ranjan) brings out the forces and factors proving stumbling blocks in the implementation of the act. It also highlights the reluctance on the part of government, most notably the role of Forest Department, which is leaving no stone unturned in nonimplementation of this act because it fears that the department will lose its hegemony over forest areas and its hold over pouring billions of rupees in the name of afforestation and creating a forest that can act as a "carbon-sink." It also highlights other factors at work, most particularly the role of mining companies and their lobbies (working both at the central and state levels), which find this act a spoiler and dampener in their designs.

The third chapter (Niraj Kumar and Mahendra Prasad Singh) presents a case study of the Telecom Regulatory Authority of India (TRAI) to showcase the conversion of "India's centralized bureaucratic state into a federal regulatory state" from a micro perspective. It highlights the significant changes in both India's mixed economy and "union executive-driven political system" brought about by economic liberalization/privatization/globalization. The central question raised here is as to how some facets of neoliberal reforms are impacting the federal state (in relation to the provincial states, civil society and the market). The neoliberal reforms in India since 1991 brought about a paradigm shift. Economic liberalization and globalization have led to the creation of a range of autonomous and civil judicial regulatory authorities in various sectors of national and state economies, replacing a great deal of direct administrative or bureaucratic control of the economy. The early parliamentary enactments such as the TRAI Act, 1997, amended in 2000, did not ensure security of tenure to the regulator. This, however, changed in subsequent acts where the security of tenure was greater, with more autonomy given. The slender power given by the government to the TRAI has led to infringement by judicial bodies on the regulator's sphere of operation, with the effect of the regulator becoming complicit in government's policies.

Furthermore, in the fourth chapter (Amna Mirza and Bhav Nath Jha), a debate is introduced on the scope and boundaries of the relationship between the Delhi government and the center, as at times both the center and Delhi government contest each other's right to administer the National Capital. This has given impetus for demands to grant statehood to Delhi. It uses the case study of the Delhi Police that reports to the union Home Ministry, which in turn restricts their arena of authority in ensuring maintenance of law and order in the capital. Parallels cannot be drawn with other state capitals, as Delhi has to represent the best of national interests. The chapter narrates the delegation of authority using yardsticks of feasibility, past reflections and necessity.

The objective of Chapter 5 (Biswajit Mohanty and Babita Verma) is to show that federalism is generally understood as layered negotiations for sharing of power among actors at multiple levels. What the mainstream theories of federalism advocate is how power and responsibility can best be shared among different levels of governments so that financial and administrative goals can be carried on efficiently. Whereas the normative economic theory of federalism vouches for "fiscal equivalence" through the successful distribution of power between different tiers of governance, land and resource federalism, in contrast, stresses land resource management through the procedural governance of "right to notice," "right to be heard" and "right to appeal." The multilevel federal governance, "the sum of regulations brought about by actors, processes as well as structures," on the other hand, although it stresses the negotiations and engagements of institutionalized actors in deliberating, formulating and implementing policies, does not take into account the movement actors who are struggling to retain their rights to common property resources and the livelihood associated with those rights. Thereupon the fifth chapter takes the actor-centered federal approach, and the case study of the anti-POSCO movement in Odisha argues that federalism as a concept as well as practice is not only about territoriality and distribution of power among units but also about "actors/agency," "spatiality," "institutional entity" and the protection of rights of people to "decide," "act" and pursue their "livelihood."

Chapter 6 (Kamal Nayan Choubey) probes as to how far the previously mentioned new legislation in the post-liberalization era has changed the structure and nature of forest governance in India and whether it would be right to claim that these laws have challenged centralization in matters related to forests and give more powers to the *gram sabha* and local communities and in this sense strengthen the third level of federal structure in the forest areas of India. From the colonial period onwards there has been an inclination toward centralized control of India's forest resources. Interestingly, in the post-liberalization era the role of private corporate capital has been immeasurably increased in the extraction of forest resources for the "development," which further dispossessed the tribal population in these areas. Since the 1970s, however, there has been a parallel process in these areas, too, and tribals have resisted the imposed and centralized development model. The resistance by the tribals has been related to the "struggle for better or more progressive laws." The parliament has enacted two such laws for forest areas. One of them is Panchayat (Extension to Scheduled Areas) Act 1996 (PESA) and the other law is Scheduled Tribes and Other Traditional Forest Dwellers (Recognition of Forest Rights Act) 2006 (or Forest Rights Act or FRA).

Chapter 7 (Avantika Singh) bridges the gap by providing an in-depth analysis of the Indian case and its experience with the climate policy dissemination through its ever-evolving federal route. An attempt to sketch the picture is made via critical comparative examination of the National Action Plan for Climate Change (NAPCC) and the State Action Plan on Climate Change of Uttar Pradesh (SAPCC). The author argues that environmental degradation originating in one place impacts on a much larger geographical area, and to counter such degradation not only is the role of local governments significant, but the interposition of state and central governments also is required. Therefore, the conception of environmental federalism entails an inspection of the apposite authority for the administration and delivery of environmental goods and services. In this the role of the central government becomes crucial with respect to the environmental regulation that necessitates taking charge for those actions that have vital "environmental spillover effects across jurisdictional boundaries."

Chapter 8 (Himanshu Roy) presents an analysis of the role of regional business, jointly signified by the trade associations and different regional chambers of commerce and industry, in the creation of an Indian federation premised on linguistic provinces in colonial India and its further federalization through fiscal and administrative reforms or political accommodation in postcolonial India. The process is still ongoing and has been in continuation for the past 125 years, since 1887, when the Indian business chambers began to be established in different parts of the country. The chapter signposts an emerging subterranean trend of gradual marginalization of regional parties that has played a significant role in the federalization process in postcolonial India. The new social milieu of globalization that propels standardized attributes of the market in different regions drove the regions closer to each other with similar traits, making the regional parties with their unique identities irrelevant.

Chapter 9 (Alisha Dhingra) examines the interaction of federalism and the welfare state in India and its changing dimensions under the impact of globalization leading to the emergence of varied welfare regimes in India. It challenges the neoliberal-welfare dichotomy, arguing that contrary to the perception of the shrinking of welfare expenditure under neoliberal reforms, India's welfare expenditure has been on the rise, although the burden of allocation has shifted from the central government to state governments. This leads to differential entitlements for citizens across different states.

Chapter 10 (Sangita Dhal) explores the role of different actors involved in implementation of ICDS and the presence/lack of cooperation among them. Furthermore, with the adoption of neoliberal policies, there has been a modification in the approach toward the development and redistribution of wealth. The state in the era of globalization has espoused "a new model of a rights-based approach of social inclusion and equitable distribution of the fruits of development." It is believed that the growth of the economy and consequent development of the nation in general are hinged upon corresponding development of the social indicators. The Integrated Child Development Scheme (ICDS) is one such social welfare program launched by India; it is the oldest nutrition programs in the world that aims at providing food, preschool education and primary health care to children up to 6 years of age and pregnant and lactating women. The center, through the implementation of the Integrated Child Development Scheme in the various states of India, seeks to fulfill its social and constitutional responsibility by adopting a proactive approach toward the poor and underprivileged. Based on an empirical study, the author gives a clear insight into the working of ICDS, especially at the state and substate levels. It also elaborates on the institutional mechanism of ICDS and evaluate its impact as a Centrally Sponsored Scheme (CSS) by exploring the process of its implementation in Jajpur district of Odisha, which is a progressive rural district with a dominance of scheduled caste population.

Chapter 11 (Rahul Mukund) features an investigation of the role street-level bureaucrats play in the implementation of rural employment policy in the Jharkhand province of India. The aim of the study is to highlight the bottlenecks in MGNREGA application and thereby engage in conceptual approaches based on empirical evidence obtained from the fieldwork in four villages of Jharkhand. The empirical observations from the fieldwork point toward three aspects: delay in delivery of wage payments, delay in procurement of construction material for building dug wells, and the contravening of the provisions mentioned in the act. Based on the empirical evidence, the role of street-level bureaucrats and their impact on addressing the problems faced by beneficiaries is discussed. In the process of examining the bottlenecks in MGNREGA implementation in Jharkhand, the functional operation at the local level comes out for the Indian state.

Conclusion: Strengthening cooperative federalism

The idea of "cooperative federalism" was first employed in comparative federal theory and politics, as opposed to the "dual federalism" usual in the 19th and early 20th centuries, in the stir of the Great Economic Depression of the 1930s and the Second World War once the emergencies required that the two levels of governments in the Anglo-American and British Commonwealth federal systems placed their jurisdictional battles behind them and functioned in a really national and cooperative way to tackle the difficulties confronting them. Following the war, cooperative federalism persisted to institute and apply welfare state policy regimes in the United Kingdom and Canada (and somewhat also in the United States, although there has been reduced interest in welfare liberalism in that country) to deliver inclusive social and economic securities to the people devastated by the economic depression and war.

There is weak institutionalization of shared rule mechanisms in India. The Rajya Sabha (second chamber of Parliament) is ineffective in representing state interests, as the states are represented in proportion to their population, and there is no residency requirement for members of the Rajya Sabha. Moreover, nomination to the Rajya Sabha hinges more on central party support instead of regional party support in the state assemblies, which officially elect the members of the second chamber. Therefore, the Rajya Sabha is more bound to party political interests than to provincial interests. Interministerial conferences have advanced in various policy areas of shared central–state interest, but the Inter-State Council did not progress into the envisioned apex body of center–state partnership. Thus, there is an indisputable need to bring together the executive heads of the two levels of the governments, that is, prime ministers and key union ministers and chief ministers of the state governments and lieutenant governors of the UTs at forums representing executive federalism, viz, the Inter-State Council and National Development Council. This will only valorize the cause of cooperative federalism through intergovernmental consultation.

Whether it is the recommendation of the Sarkaria Commission or of the National Commission for the Review of the Working of the Constitution (NCRWC) (under the chairmanship of Justice M.N. Venkatachaliah) or of the Second Commission on Centre State Relations (under the chairmanship of Justice M. Punchhi), all endorse the empowerment of the intergovernmental forums like ISC and NDC and activating them by making them the institutionalized mechanism of union–state consultation and decision making in matters of domestic as well as foreign policies.

Swenden and Saxena critically assess the impact of the Planning Commission on center–state relations in India. They argue that the Planning Commission had a centralizing effect in performing its role of administration of five-year and annual planning and its involvement in scheming and managing Centrally Sponsored Schemes (CSS) as well as its contribution in discretionary grant making. The policy primacies of central government and interstate divergences prevented the Planning Commission from gaining the character of a shared rule institution, proficient in counteracting the centralizing tendencies (Swenden and Saxena 2017). The Planning Commission was regarded as a "de facto arm of the central government" that, through its involvement in grant making and central and state planning, supplemented the centralization of Indian federalism. Moreover, the Planning Commission also seemed to corrode some of the actions of the Finance Commission, (a statutory body that has allocated pooled tax revenues to the center and the states in a less discretionary and more formulaic manner) (Swenden and Saxena 2017).

A major structural reform in this regard is the transformation of the Planning Commission to NITI Aayog (National Institution for Transforming India) set up by cabinet resolution. This policy think tank consists of a civil servant as the chief executive officer, two full-time experts (an economist and a defense research and development expert), six union minsters (three ex-officio and three special invitees), free market economist Arvind Panagariya as deputy chair and the prime minister as the chair. Its purpose is to advance a "national agenda" to be used by the prime minister and the chief ministers to promote "cooperative federalism." Its Governing Council includes all the chief ministers of the states and lieutenant governors of the union territories, again with the prime minister as chair (Tewari and Saxena 2017).

The ISC could have been made the governing council of the NITI, but it has remained relegated by all governments, including the Janata Dal-led National Front Government, which created it for political flexibility. However, NITI as well as its Governing Council is relegated to a greater amount of informality

> as ad hoc deliberative bodies activated by referring a matter to it rather than a regular channel of advice for policy making in the government. They are thus a far cry from Nehru's Planning Commission and the NDC which was brought into existence for offering guidelines for five-year plans and approval of draft five-year plans. (Singh, Saxena and Bhardwaj 2015)

However, NITI departs from the Planning Commission in the sense that it has set up regional councils "to address specific contingencies impacting more than one state or a region, convened and chaired by the PM but made up [of a group of] Chief Ministers of the State." (Cabinet resolution January 1, 2015).

The NITI Governing Council, at its first meeting in February 2015, constituted three such regional councils concentrating on the restructuring of CSS (as reported earlier), skills development and the Swachh Bharat Abhiyan or the Clean India Mission. NITI has also set up smaller task forces, chiefly at the level of the senior civil service, with the involvement of consultants to concentrate on poverty elimination and agricultural development (NITI Brief 1. Accessible via www.niti.gov.in, accessed in April 2015).

Many experts welcomed the organizational changes in the composition of the NITI, as this would possibly give a greater involvement of the states at an initial stage in the policy formulation (Interview with Indira Rajaraman, March 9, 2015; with Amitabh Pande, March 16, 2015).

Swenden and Saxena (2017: 59) observe:

> Although it is much too early to tell whether the NITI approaches center-state interactions in a substantially different way from the PC before, the creation in 2015 of several Regional Councils under its helm appears to have given (some) states a stronger footing in the early stages of the central policy process. However, the agenda-setting capacity of the central government and in particular, the PMO and Finance Ministry's ability to determine the scope and input of the states appears to have weakened the aspirations of the NITI as a key

> intergovernmental body tasked with streamlining India's collaborative federalism. Furthermore, the continued near-absence of state-based representations among the NITI's cadre confirm its status as more of a central than intergovernmental think-thank.

Swenden argues that in comparison with the Planning Commission, the NITI has been able to offer more short-term contracts to externally recruited policy analysts, but mostly at the junior level. He believes that the NITI could have fetched in civil servants or practitioners from the states in order to realize its role as a "federal" think tank or intergovernmental hub, but largely failed to do so. He argues that in terms of working with the states in Team India and nurturing "competitive cooperative federalism," the contribution of the NITI is mixed at best. In terms of its organizational structure, the NITI is a central political institution subsumed under the authority of the prime minister, who serves as the NITI's chairperson; other members include a CEO with the rank of secretary, a vice chairperson appointed by the PM, three fulltime members, four ex-officio members and three special invitees (all belonging to the central ministers). One of the member of the NITI Aayog emphasized that "as the NITI reports to the PM, it is logical for the PM to choose who will become its members and to set the NITI's agenda" (Swenden 2019).

Another important change in structure of tax reforms is the transition from the Value Added Tax (VAT) to the Goods and Services Tax (GST), which ensured making India a common national market, with a GST Council for its control comprising the union finance minister as the chair and state finance ministers as members, with two-thirds vote for the states block. Nonetheless, a majority decision in the GST Council must include the union finance minister. Specialists argue that the GST would boost the "ease of doing business in India" and expand revenues of governments. But for the consumers, goods would get cheaper and services including education and health care costlier. The reality is that incorporating petroleum products, electricity, real estate and alcohol into the list of exemptions limits the scope of the common national market. Further, a general Trade and Commerce Commission imagined by the framers of the Constitution is elusive.

Further, critics like A. Navaneethakrishnan observe that

> there are no clear-cut guidelines to the GST Council regarding how to share the taxes. The composition of council is political. In the absence of guidelines, how the GST Council will share the tax. There is no formula regarding this. Therefore, I would like to submit that the GST bill is unconstitutional.

Pinaki Chakraborty from the National Institute of Public Finance and Policy (NIPFP) associates the voting pattern within the council to the loss of fiscal autonomy of states, as she says, "since any change would mean 3/4th of the total votes, and union government accounting for one third of the voting share, without union government's approval changes cannot be carried out" (Select Committee Report 2015).

Today the leading party and its leaders, often with unreasonable and unseemly vengeance, have learnt the importance of consultative coalition governance, thanks to the pressure system created by coalition governments. However, the opposition state governments are still treated with unconstitutional impropriety and injustice by the ruling arrangements at the center. Intergovernmental relations based on constitutional norms, values and rules need to be promoted expediently.

Apart from the sphere of the economy, the center and state governments have collaborated to address security concerns such as Maoism. Operation Green Hunt 2009 was an all-out offensive by paramilitary forces and the state forces along the borders of Chhattisgarh, Jharkhand, Andhra Pradesh and Maharashtra. The problem was the lack of clarity on the range of troop deployment, the configuration of the forces and the chain of command between central paramilitary forces and the state police.

The cooperation between the center and states has also been a constant concern in the social sector, prominently in the sphere of education and health. The 42nd Amendment of the Constitution of India in 1976 transferred the subject of education from the state list to the concurrent list, implying concurrent (greater) jurisdiction of the union in education, which previously was exclusively under state jurisdiction. Besides the constitutional amendment, various acts of the union and many subordinate legislations of the union in the field of education besides significant policy interventions have been instrumental in reshaping the federal space in education. The proliferation of central regulating agencies and Centrally Sponsored Schemes in education are significant developments that have added new and important dimensions to union–state relations in India. In the backdrop of post-constitutional developments in the field of educational federalism, it would be interesting to map out the changing dimensions of union–state relations in education. The policy initiatives of the union in recent decades, including the establishment of central regulating agencies and the recently launched Centrally Sponsored Scheme of RUSA will form the basis of understanding union–state relations.

The issue of health assumes utmost importance in the event of pandemics. The spread of COVID-19 across the world has emphasized the need for cooperative federalism to deal with pandemics, since all levels of governments need to adopt measures to curb and manage the pandemics. In most countries across the world, health care is the responsibility of provincial governments, but during pandemics, when the life of citizens is at stake, the central governments need to take a proactive role. Local governments also need to take preventive measures and cope with the fallout of the other levels' intervention measures. Thus, regardless of the distribution of powers and responsibilities, all the levels of government need to work in close coordination with each other, signifying the spirit of cooperative federalism. Thus, despite the measures taken by the central government applicable uniformly across the country, some states might adopt some innovative ways of dealing with the crisis, and other states might learn from them. But if state governments, instead of cooperating and learning from each other and coordinating with the central government, start obstructing the measures taken by the central government, this might lead to intensification of crises. Thus, if all levels of government work in coordination and harmony and learn from each other, the federal system would become a boon, but the lack of harmony and cooperation will make it a bane in the event of a pandemic and other crises situations (Styetler, Mail Communication, March 24, 2020, at 12:55 AM).

Another example of cooperative federalism is the announcement by the prime minister in his Independence Day Speech of Ayushman Bharat, the National Health Protection Mission (AB-NHPM) or Modicare, launched on 25 September 2018. Under the Pradhan Mantri Jan Arogya Abhiyaan, 10 crores families will get Rs 5 lakh health insurance per family per year. This scheme covers over 1 million poor households, or 50 million people, based on the socioeconomic caste census, for secondary and tertiary care hospitalization. Non-BJP ruled states like Delhi, West Bengal, Punjab and Odisha criticized the scheme, as they believed that the scheme is sparse and they would rather continue with their "better" health insurance schemes, which will cover more people as compared to Modicare. Though the state is supposed to pay 40 percent of the total funds whereas the center will support 60 percent, these states are of the opinion that they will spend roughly the same amount or even less in their own health care schemes and get full credit for launching it. Since health is a state subject, it might become a more complicated issue in the years to come.

Since 2014, the BJP-led NDA government is promoting the agenda of cooperative federalism. The establishment of the NITI Aayog and the GST Council have been argued by the government to be steps in the direction of cooperative federalism. However, critics consider the agenda of cooperative federalism pursued by the government as political rhetoric, as the domination of the central government is visible and the confrontation between the central government and the state governments has increased substantially after the 2019 elections, in which the BJP-led NDA secured a thumping majority in Lok Sabha.

Indian federalism is at the crossroads. It is the dialectics of competition and collaboration that will define the direction of federalism in India. Although the competition among the states and between the center and states has increased in the era of globalization, the challenges posed by globalization in different areas such as security, the social sector and foreign relations require more and more collaboration between different levels of governments. Collaboration at both the vertical and horizontal levels is the hallmark of a federal system. Such a collaboration becomes essential to meet security threats. These threats may be posed by natural disasters like the recent flood in Kerala. The center and the states deal with such calamities together. Prime Minister Narendra Modi announced a grant of 500 crores apart from the 100 crores declared by Home Minister Rajnath Singh. Moreover, Delhi Chief Minister Arvind Kejriwal has announced a 100 crore-relief fund for Kerala. Another example of horizontal collaboration was the setting up of a centralized secretariat at Panchkula to jointly confront the drug menace in the northern region of the country. The centralized secretariat has been set up by the six states (Haryana, Punjab, Himachal Pradesh, Rajasthan, Delhi and Uttarakhand) and the union territory of Chandigarh, where nodal officers from each state will be deputed for the purpose of sharing intelligence and information. Thus, we can see both vertical and horizontal collaboration at work in these cases. These could provide paths toward strengthening cooperative federalism in India.

References

Basu, Partha Pratim. 2016. 'Federalism and Foreign Policy in India: Experiences of UPA and NDA-II Regimes', *India Quarterly: A Journal of International Affairs*, Vol. 172, No. 3. https://doi.org/10.1177/0974928416654367, accessed on 24 July 2018.

Bhattacharya, Harihar and Lion König. 2016. 'Conclusion: India's Second "Tryst With Destiny"', in Harihar Bhattacharyya and Lion König (eds), *Globalisation and Governance in India: New Challenges to Society and Institutions*, London and New York: Routledge.

Cabinet Resolution. 1 January 2015. 'Interview with Indira Rajaraman, Former Member of the XIII Finance Commission, Delhi, 9 March 2015', The Gazette of India, REGD NO L.-33004/99.

deSouza Peter Ronald. 2003. 'The Struggle for Local Government: Indian Democracy's New Phase', *Publius*, Vol. 33, No. 4, pp. 99–118.

Dhavan, Rajeev and Rekha Saxena. 2004. 'The Rewriting of Indian Federalism: Constitutional Amendments, Statutory Changes and the Executive Power Revisited', Typescript PILSARC.

Dua, B.D. and M.P. Singh. 2003. *India's Federalism in the New Millennium*, Delhi: Manohar Publications.

Govinda, Rao M. 2006. 'Fiscal Federalism in India: Emerging Challenges', in Raoul Blindenbacher and Abigail Ostien Karos (eds), *Dialogues on the Practice of Fiscal Federalism: Comparative Perspectives, a Global Dialogue on Federalism*, Booklet Series 4, Forum of Federations and ISCFS, Ottawa.

Interview with Amitabh Pande, 16 March 2015.

Jenkins, Robert. 1995. 'Theorising the Politics of Economic Adjustment: Lessons From the Indian Case', *The Journal of Commonwealth and Comparative Politics*, Vol. 33, No. 1, pp. 1–24.

NITI Brief 1. Accessible via www.niti.gov.in, accessed on April 2015.

Pai, Sudha, ed. 2013. *Handbook of Politics in Indian States: Regions, Parties and Economic Reforms*, New Delhi: Oxford University Press.

Report of the Commission on Centre-State Relations headed by M.M. Punchi, March 2010.

Rudolph, Lloyd I. and Susanne Hoeber Rudolph. 2001. 'Redoing the Constitutional Design: From an Interventionist to a Regulatory State', in Atul Kohli (ed), *The Success of India's Democracy*, Cambridge: Cambridge University Press.

Saez, Lawrence. 2002. *Federalism without a Centre: The Impact of Political and Economic Reform on India's Federal System*, New Delhi: Sage Publications.

Saxena, Rekha. 6 January 2007. 'Treaty-Making Powers: A Case for "Federalisation" and "Parliamentarisation"', *Economic and Political Weekly*, Vol. 42, No. 1, pp. 24–28.

Saxena, Rekha. 2012c. 'NCTC Can't Work without States Support', *The Pioneer*, 18 April.

Saxena, Rekha. 2012d. 'Lok Pal Bill Remains Hostage to Political Uncertainties', *The Pioneer*, 3 April 2012.

Saxena, Rekha. 14 January 2012a. 'Is India a Case of Asymmetrical Federalism', *Economic and Political Weekly*, Vol. 47, No. 2.

Saxena, Rekha. 2012b. 'Centre Can't Push the States Around', *The Pioneer*, 25 April.

Select Committee Report on 'The Constitution: (One Hundred and Twenty-Second Amendment) Bill, 2014', Rajya Sabha Secretariat, New Delhi, 22 July 2015.

Singh, M.P. and Rekha Saxena. 2011. *Indian Politics: Constitutional Foundations and Institutional Functioning*, New Delhi: PHI.

Singh, M.P. and Rekha Saxena. 2013. *Federalizing India in the Age of Globalization*, New Delhi: Primus Books.

Singh, M.P., Rekha Saxena and Ritesh Bhardwaj. January–June 2015. 'Modi's "Cooperative Federalism": A Policy Promise in Search of Implementation', *Half Yearly Journal of Bihar Political Science Association*, Vol. 4, No.1, pp. 20–46.

Sinha, Assema. 2004 January. 'The Changing Political Economy of Federalism in India: A Historical Institutionalist Approach', *India Review*, Vol. 3, No. 1, pp. 25–63.

Styetler, Nico. 2020. Mail Communication, 24 March 2020 at 12:55 AM.

Swenden, Wilfried. 2014. Concept Note for Leverhulme International Research Network Project Titled 'Continuity and Change in Indian Federalism'.

Swenden, Wilfried. 2019. '"Team India" and the NITI Aayog', *Seminar*, Issue 717, May 2019.

Swenden, Wilfried and Rekha Saxena. 2017. 'Rethinking Central Planning: A Federal Critique of the Planning Commission', *India Review: Continuity and Change in Contemporary Indian Federalism*, Vol. 16, No. 1, pp. 42–65.

Tewari, Manish and Rekha Saxena. 2017. 'The Supreme Court and the Rise of Judicial Power and the Protection of Federalism', in Nicholas Aroney and John Kincaid (eds), *Courts in Federal Countries: Federalists or Unitarists?*, Toronto: University of Toronto Press: 223–255.

Tillin Louise,"Asymmetric Federalism" in The Oxford Handbook of The Indian Constitution, eds. Sujit Choudhary, Madhav Khosla, and Pratap Bhanu Mehta. Oxford, Oxford University Press, 2016.

Part I

ADMINISTRATIVE FEDERALISM

1

ARTICLE 356 AND ITS DYSFUNCTIONS

Analysis of its misuse in the Indian federal system

Subhendu Ranjan Raj

In the constitutionally mandated federal structure of India, Article 356 has created bitter acrimony and far-reaching implications in the federal firmament. In the last seven decades or so (as of April 2019), there have been some 115 amendments to the Constitution of India since it came into force, and many of them have impinged on center–state relations. Some of such changes brought about by these amendments have impinged on the role of Article 356, as has been the pronouncement in various Supreme Court and High Court cases. Therefore, the operation of Article 356 in the Indian polity over the years is responsible for creating many imbalances in the federal architecture of India.

The focus of this chapter is to study the concatenations associated with this article and its operation and also to examine what aspects of it have led to controversies and conflicts. It needs to be pointed that Article 356 *ipso facto* has not been the source of such controversies but rather the political and federal climate as also the social and regional change paradigms in which the said article may be said to have its operationalization. An additional perspective also needs to be brought in. This may be defined as follows: namely, that responsibility for change in the federal construct has also to be diagnosed in the changes brought about by the neoliberal/globalization economy, the upsurge of dalit identity and the assertion of regionalism with the growth of lower caste parties. Hence, this chapter posits that though political manipulation of provisions of Article 356 are crucial for its misuse – despite the attempts of the constitutional courts (the Supreme and High Courts), to rein in the political misuse to some extent that have been highlighted in the recent cases concerning Arunachal and Uttarakhand, where the courts have come down heavily on the political dispensation/center – there is also a need to shed the colonial mindset or perspective with which Article 356 was constructed (as it is a legacy of Government of India

Act, 1935), and the provisions giving explicit dominance to the center's will and intent have to be watered down in the light of federal accommodation. Some modifications to this article are also due in the light of the recent pronouncements in the Arunachal and Uttarakhand (as also the *Bommai*) cases. It is also crucial that the requirements of the fiercely contested electoral shortcomings of majoritarianism and coalitional politics should not be ventilated through use and misuse of constitutional federal provisions such as Article 356 to settle political scores and destabilize legitimately selected governments. The art of federal coexistence and power sharing demands tolerance and multilevel cooperation.

The Indian federal system

It is true that the arena of Indian federalism has been a contentious one. Our Constitution makers envisioned that in any multilevel federal system of governance, like that of India, nonantagonistic relations between the center and the states are paramount for the stability, security and economic development of the country. The Constituent Assembly integrated democracy and federalism well, but there was a slant toward provision for a stronger center. Many political historians, including R.C.S. Sarkar (1986), have observed that this trend of over-centralization of powers at the hands of the union government was a major reason for the ouster of the Congress and brought on the era of contentious federal power tussle and coalitionalism in the center and the states in the post-1977 period.

With the rise of coalitions there have been a surge of anti-Congress governments in the states. Demand for greater administrative and fiscal autonomy from the center for the states increased. The political climate heated up, and center–state relations were worsening. This was noted by the Administrative Reforms Commission (1969), which had recommended that steps be taken for maintaining harmonious center–state relations. The Rajamannar Committee (1971) had prepared a case for greater autonomy for the states, basically in legislative and fiscal matters. State governments of West Bengal, Tamil Nadu and North-East, the Shiromani Akali Dal (Punjab) and the Telugu Desam Party (Andhra Pradesh) were wanting areview of the overall constitutional schema of center–state relations. This situation prompted the setting up in 1983 of the first Commission on Centre-State Relations, headed by Justice R.S. Sarkaria, a retired judge of the Supreme Court of India. The Sarkaria Commission, as it was popularly known, submitted its report in 1988. It did admirable work, but its recommendations were put on the backburner by successive governments.

The '90s were marked by an economic crisis that brought on the neoliberalization drive by the Narasimha Rao government, which actually started out as a minority government. By the logic of decentralization of economic planning and decontrolling the economy, a lot of excessive control mechanisms of the government underwent a change. States wanted more space and say in policy making and decentralization in the implementation process, especially in the sectors of public distribution, health and education. Hence, the total picture of resisting the control of the center gelled very well in the new scenario. While liberalization and giving seminal status to the market increased growth, regional disparities among the states kept on increasing. The rise of the new forces of regional parties and coalitions – as first noted by Christophe Jaffrelot in 2000 and edified in his later book *India's Silent Revolution* (Jaffrelot 2003) – in the background created the conditions for agitational politics of the states with the center. Political coalitions at the center that had started from 1977 had in the meanwhile become the *de rigeur* of Indian politics, and the first substantial coalitional governments lasting full tenures after a spate of coalitions were the National Democratic Alliance (October 1999–May 2004), followed by two terms of the United Progressive Alliance from 2004 till 2014 and the latter two terms of the NDA II alliance (from 2014–2019 and 2019–continuing).

It is against this background that our study of the use of Article 356 will be examined. Over the years, at least three types of situations have emerged in the federal context in India, which have seen the proliferation of the use of Article 356 and the resultant destabilisation of state governments and should be the subject of detailed analyses:-

i Homogeneity of political color/ideology – When the same party rules in the state/states and at the center, then center–state relations tend to be cordial and nonadversarial. The relations of the governor and president/PM/ministries of the Government of India are amicable and noncontroversial. A formalized system operates with vigor; financial aid/grants in aid do not become flashpoints of bitterness and are issues on which there is mutual give and take.
ii When state parties are part constituents of the federal coalition and the government is controlled by a state party that also participates in the central government, bilateral relations between the state and center are more formalized, as there is some iota of commonality between the state party and central ministries. Federal relations are on the upswing.
iii Heterogeneity of political color/ideology – When different party/ies coalitions rule at the center and single-party majorities or coalitions constituted by one regional party in predominance are formed in the states, then adversarial relations follow, and there are tussles between the states and the center on innumerable issues, especially on fiscal issues, the role of governor and the use of Articles 200, 249 and 356, among others. Federal relations become the arena for contestation and bickering.

The Constitution, vide Articles 200 and 201, lays down the modality of amendment by the state legislatures. A bill that is to be passed by the state legislative assembly has to be presented also in the legislative council of the state (if it has one). Thereafter, the governor either assents to it or reserves it for the president's consideration. He may also cause reconsideration of the said bill, in which case he sends it back to the state legislative assembly or council. Two eventualities may occur in this context. Thus, under Article 200 the governor may act on his discretion and decide to return any bill, other than a money bill, to the state legislature with a message. Under Article 201 governor can act independently and reserve a bill for consideration by the president.

In reality, however such constitutional requirements are not adhered to. For example, Governor Ram Naresh Yadav of Madhya Pradesh in August 2013 withheld his assent to the appropriation bill (a money bill) which covered the Third supplementary grants. He was not entitled to do this. Although he did not issue sanctions to prorogue the House, he did prorogue it. By his action in both of these matters, the governor violated the Constitution.

The Constitution unambiguously says that the states have powers to deal with items included in the state list, yet it qualifies the same by stating that under Articles 249, 250, 252 and 253 there are eventualities when the federal government may legislate on these items. Of these, in particular, Articles 248 and 249 grant the residuary power of legislation to Parliament. The Constitution grants exclusive power to Parliament to make law on any subject matter not covered by the state or concurrent lists. Under Article 249 of the Constitution, Parliament is also empowered to legislate in the "national interest" on matters covered by the state list. In case there is any inconsistency between the law made by Parliament under Article 249 and a law made by a state legislature, the law made by Parliament shall reign supreme. Further, Parliament can enact laws on state subjects for those states whose legislatures have given their consent to such central legislation.

Regarding the use of Article 356, eminent jurist Soli Sorabjee is of the view that after the *S. R. Bommai* judgment, it is quite well settled that Article 356 gives supreme power and is to be applied only when the constitutional machinery has broken down irretrievably (Reddy and Joseph 2004). The Sarkaria Commission Report had mentioned way back in 1987 that even though many governments enjoyed majority support in their respective legislatures, the center imposed Emergency under this article in 13 states. There were instances where Article 356 was imposed to deal with intra-party problems or for reasons for which the provisions of this article are not relevant. This author has found that, in actuality, President's rule has been imposed about 119 times (through December 2019), and it covers almost all the states of the Indian Union under different pretexts and political contexts.

Article 356: its provisions

In this section we focus on the provisions of Article 356 and make a detailed analysis of its usage in India subsequently. It may be pointed out that Part XVIII of Constitution lays down the following provisions in this regard:

a Articles 352, 353,354, 358 and 359 deal with emergency in the union.
b Articles 356, 357 and 358 deal with the imposition of emergency in the states (known in popular nomenclature as President's rule) in case of the emergence of certain situations.
c Article 360 deals with a financial emergency in the country.

Article 356 (1), as outlined in the Constitution, lays down the following:

> If the President, on receipt of a report from the Governor of a State or otherwise, is satisfied that a situation has arisen in which the Government of the State cannot be carried on in accordance with the provisions of this Constitution, the President may by Proclamation:
>
> (a) assume to himself all or any of the functions of the Government of the State and all or any of the powers vested in or exercisable by the Governor or any body or authority in the State other than the Legislature of the State;
> (b) declare that the powers of the Legislature of the State shall be exercisable by or under the authority of Parliament;
> (c) make such incidental and consequential provisions as appear to the President to be necessary or desirable for giving effect to the objects of the Proclamation, including provisions for suspending in whole or in part the operation of any provisions of this Constitution relating to anybody or authority in the State.[1]

Thus, under Article 356, if the

> President is satisfied that such a situation has arisen, whether on the basis of a report received by the governor of a state or otherwise he may by proclamation take any or all the three steps mentioned in sub clause (a), (b) and (c).

It is to be noted that under Article 355 that the union has the responsibility to ensure that the working of the governments in the states in accordance with the provisions of the Constitution. It may be said that the US Constitution (Article IV.4) and the Australian Constitution (section 61) have a similar provision. Though emulated in some measure from these sources, the real inspiration for inclusion of the constitutional provision contained in Article. 356 lies elsewhere, as will be discussed in the next segment.

Its history

The contentious article, namely Article 356, which is the source of innumerable controversies, is inspired by the Government of India Act, 1935, of colonial times. Section 93 of this act provided that

> if a Governor of a province was satisfied that a situation has arisen in which the government of the province cannot be carried on in accordance with the provisions of the said Act, he could assume to himself all or any of the powers vested in the provincial body.

Although Article 356 is modeled on this provision, there is a slight modification – namely, that instead of the governor, the president is vested with such power.

It has been argued by many both within the Constituent Assembly and outside that only a spirit of "cooperative federalism" can tangibly preserve the balance between the union and the states and promote the good of the people, not an attitude of dominance by the party in power at the center. The power of governance is distributed in many organs and institutions of the government in the multilevel federal system that we have in India. So a consensus has been reached (something which has been reiterated in the many court cases in this article) that the extraordinary power conferred in Article 356 should not be employed to further the prospects of any political party or to destabilize any elected government or a duly constituted legislative assembly. The consequences of such misuse will be deleterious for the polity in the long run. It has already proven to be so. This has been in large measure because the spirit of cooperative federalism had evaporated in real practice of the functioning of the federal system in India, long hence.

An article like 356 is not be found in any democratic constitution across the world except the constitutions of Indian and Pakistan, both sharing the historical background of the Government of India, 1935 Act. Thus the expectation that Article 356 would be invoked in extreme situations and will not be utilized as a "surgical operation for a mere cold or catarrh"[2] and the hope of the Constituent Assembly that this provision "will never be called into operation" and "would remain a dead letter"[3] has not happened in our federal relations of the past seven decades.

Article 356: as it stands now

History is witness to the fact that the Article 356 has been invoked some 119 times to date (i.e., December 2019). All political parties, be it the Congress,

Bharatiya Janata Party or even the coalitional outfits such as the United Progressive Alliance (UPA) and the National Democratic Alliance (NDA), have misused the power under Article 356 when they have been in power at the center on many occasions. *Sarkaria Commission Report* (Chapter Six) and the *S.R. Bommai vs UOI* case (AIR 1994 SC 1918) have categorically substantiated that on many occasions *state governments have been pulled down even though they had a legislative majority, or in certain cases state governments were disposed of without giving them a chance to prove their legislative majority* on the floor of the Assembly. In the path-breaking *Bommai* case, despite requesting to allow Bommai the opportunity to prove his party's majority in the Karnataka Legislative Assembly on the floor of the house within a short span of time, the governor denied him that opportunity and went on to dismiss his ministry. This action found strong condemnation in the Supreme Court.

The use and misuse of Article 356

The Ministry of Home Affairs of the Government of India informs[4] that Article 356 had been imposed 115 times through June 2018. At the time of writing, this has spiked up to 119, the last instance being in Maharashtra in November 2019.

Figure 1.1 gives decadal data regarding the imposition of Article 356 in various states of the Indian Union. It categorically highlights that the period of the '70s through the early '90s were a major example of political turmoil. The Congress Party was undergoing a phenomenal change from being the "umbrella party," which some commentators such as James Manor and Rajni Kothari have commented upon. The Congress party used Article 356 around 88 times during the 54 years they held power in the center. Jawaharlal Nehru, the first prime minister from this party, used Article 356 some eight times. Lal Bahadur Shastri, his successor, used it just once, in Kerala in September 1964. Indira Gandhi, however, beat them both, as she holds the record for invoking Article 356 the maximum times during Congress rule, that is, 50 (35 times in her first phase from 1966 to 1977 and 15 times in her second stint from 1980 till 1984). The reason was that Indira Gandhi's ascendancy post-Shastri brought into focus a sharp political divide with the states that had opposition parties in power. This divide increased after her reinstatement in 1980 after her unceremonious exit. She put opposition governments out of power by whimsically using Article 356, both in her pre-1977 and post-1980 reign. For example, in April 1977, Bihar Jagannath Mishra's government was dismissed and Article 356 imposed.

Similarly, in April 1977, Shyama Charan Shukla's government was disposed of despite having majority support in Madhya Pradesh Assembly. In February 1980, the Gujarat government of Babubhai J. Patel was dismissed despite having a majority in the House. In the same month in Madhya

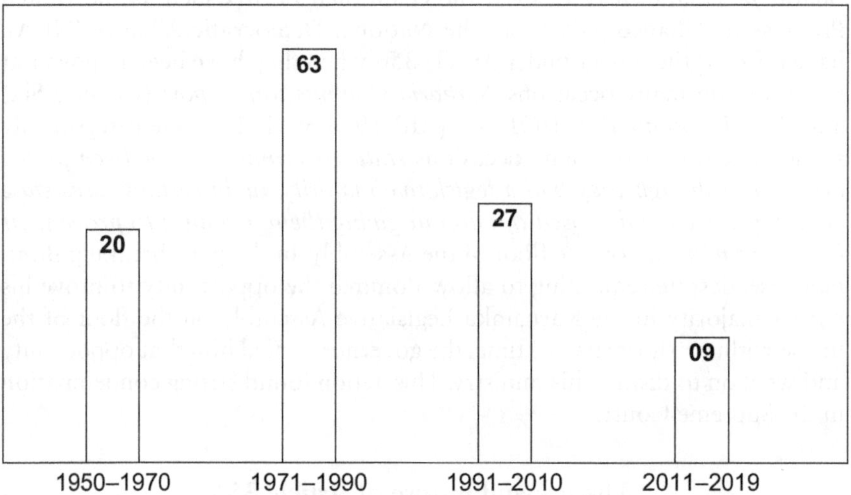

Figure 1.1 Decadal Instances of Imposition of President's Rule in States Since 1950
Source: Compiled by the author from various sources

Pradesh, Sundar Lal Patwa's government was pulled out despite having a majority in Assembly. In Rajasthan, two cases occurred in April 1977 and in February 1980, when first Hari Dev Joshi and later Bhairon Singh Shekhawat, as chief ministers, were forced to lose power. They had sizable majorities at the time of their dismissal.

In Table 1.1, a consolidated figure of the total number of times Article 356 has been imposed on governments of different political parties/combinations that have been in power right after independence is given. Table 1.2 shows the break-up of the frequency of such imposition according to the prime ministers/political coalitions in power. These two tables show a spike that manifestly proves that the period just before the *Bommai* judgment of the Supreme Court in 1994 was a tumultuous period in India's federal relations. Between 1971 and 1990, the article was imposed 63 times. Later, when NDA and UPA alliance phases came into vogue, such an imposition of President's rule took place 36 times.

It ought to be noted that after the landmark *Bommai* case by the Supreme Court in 1994, Article 356 has come to be used rather sparingly, with the court issuing strict guidelines regarding its imposition. The United Progressive Alliance (UPA) led by Congress invoked President's rule in only 12 instances. The NDA government that followed it has used it in five instances to date in April 2019 (but some impositions, as we will discuss in a later section, were very controversial and contentious, on which the High Court and Supreme Court were approached, and the courts have taken a proactive stance). Under

Table 1.2 Use of Article 356 by Parties/Coalitions in Power at the Center (1950–April 2019)

Use of Article 356 by Parties/Coalitions in Power at the Centre (From 1950–April 2019)	
Congress Party (Nehru, Shastri, Indira, R. Gandhi, N. Rao)	84 times
Janata Alliance (under Morarji Desai, V. P. Singh)	18 times
Charan Singh	4 times
H. D. Deve Gowda	1 time
Chandrasekhar	5 times
UPA I &II (under Manmohan Singh)	12 times
NDA (under A. B. Vajpayee)	5 times
NDA (under Narendra Modi)	6 times*

Source: Compiled by the author from following sources: https://factly.in/state-presidents-rule-number-times/;www.livemint.com/Politics/SJ3mETZ7H1cjKNlodkcM8O/How-Presidents-Rule-in-India-has-been-imposed-over-the-year.html (both accessed on 3 September 2018) and www.indiatoday.in/india/story/maharashtra-put-under-presidents-rule-ncp-congress-bjp-shiv-sena-fail-1618224-2019-11-12

*Includes latest instances of President's rule imposed in J&K on 18 December2018 (it may be noted that Governor's rule was imposed in Jammu and Kashmir on 19 June 2018 when Chief Minister Mehbooba Mufti's coalition fell; a spell of six months of Governor's rule was declared in the state, as is required under the separate constitution that J&K has –under Article 92 of the separate constitution, that the state will be governed by President's rule can be declared subsequent to six months of Governor's rule) and in Maharashtra on 13 November 2019 (Maharashtra became the first case where Article 356 was imposed because of supposed inability of political parties to form a government after assembly elections).

NDA-II under Modi, it was first imposed in Maharashtra (September–October 2014), followed by Arunachal Pradesh (January–February 2016); twice in Uttarakhand (from 27 March 2016–21 April 2016 and 22 April 2016–11 May 2016) and in J&K from 19 December 2018. (It may be noted that emergency was imposed in J&K in accordance with a different procedure, as laid down by the separate J&K constitution. First, Governor's rule was imposed on 19 June 2018 after Chief Minister Mehbooba Mufti resigned following the loss of her coalition partner. The state was under Governor's rule for six months. It is to be noted that only after a spell of six months of Governor's rule was declared in the state, as is required under the separate constitution, that Article 356 can be imposed in J&K. Under Article 92 of the separate constitution that the state is governed by, President's rule can be declared subsequent to six months of Governor's rule. Accordingly, Article 356 was imposed on 19 December 2018 in J&K.)

It is clear that such has been the frequency of President's rule that, as of today, it has been imposed in 27 out of the 29 states of the country. Only Telangana and Chhattisgarh, the two new states that were formed recently, have remained free of such imposition. This means that in all the remaining 27 states, President's rule has been imposed. U.P. leads the pack with 10 such

Table 1.3 Frequency of Imposition of Article 356 in States in the Terms of Successive Prime Ministers

Prime Minister	*Term*	*Article 356 imposed (no. of times)*
Jawaharlal Nehru	Aug. 1947–May 1964	8
Lal Bahadur Shastri	June 1964–Jan 1966	1
Indira Gandhi	Jan. 1966–Mar. 1977	35
Morarji Desai	Mar. 1977–June 1979	16
Charan Singh	July 1979–Jan. 1980	4
Indira Gandhi	Jan. 1980–Oct. 1984	15
Rajiv Gandhi	Oct. 1984–Dec. 1989	6
VP Singh	Dec. 1989–Nov. 1990	2
Chandrasekhar	Nov. 1990–June 1991	5
P V Narsimha Rao	June1991–May 1996	11
H D Deve Gowda	June 1996–April 1997	1
A B Vajpayee	Mar. 1999–May 2004	5
Manmohan Singh	May 2004–2014	12
Narendra Modi	May 2014–Dec. 2019	6

Source: Compiled by the author from news reports; www.livemint.com/Politics/SJ3mETZ7H1cjKNlodkcM8O/How-Presidents-Rule-in-India-has-been-imposed-over-the-year.html (accessed on 3 Dec 2018) and www.indiatoday.in/india/story/maharashtra-put-under-presidents-rule-ncp-congress-bjp-shiv-sena-fail-1618224-2019-11-12

impositions at various points in time, followed by Bihar with nine impositions. In Kerala, Manipur, Odisha and Punjab it was imposed eight times each. Gujarat and Karnataka have had five impositions each. In Uttarakhand, Mizoram and Arunachal Pradesh, President's rule has been imposed only once so far.

In terms of duration it is the state of Punjab that has been put under Article 356 for the longest duration in independent India – close to ten years due to militancy in the state. Punjab had President's rule for 3,510 days (was under it continuously for five years from 1987 to 1992 due to militancy in the state). J&K comes a close second with 2,061 days – this figure will increase with the recent imposition in December 2018. In fact, both in Punjab and J&K, President's rule exceeded the constitutionally mandated three year tenure for this imposition at a stretch.

The facts cited in this section show how the federal construct has been subverted by successive parties and alliances/coalitions in power. Anil Sadanandan says correctly that

> Perhaps the two most egregious instances of mass imposition of President's rule as documented by Sarkaria Commission in 1977 and 1980 when the Janata party and Congress party respectively swept to power and dismissed opposition governments en masse. In at least 13 further instances from independence to 1986, state govern-

> ments that enjoyed a majority were dismissed. In a dozen more, no opportunity was given to claimants to prove their support, a practice declared illegal by the landmark *Bommai* judgement of 1994.
>
> (Rukmini, Bansal and AMP 2016)

Statistical inferences from India's politico-federal history

It is widely known, as we made a reference to in an earlier section in this chapter, that the Sarkaria Commission (1988) recommended the appointment of governors from an elite and well-educated/experienced background, namely from a panel of "eminent people from non-political backgrounds," in order to improve center–state relations. This has been conveniently been overlooked by successive governments coming to power over the years.

Tables 1.4 and 1.5 posit a statistical analysis of the entire gamut of history of some 65 years of gubernatorial history.

Table 1.4 shows that an overwhelming majority of governors, that is, almost 78 percent (52.3 percent and 25.6 percent were from among politicians or from the bureaucracy, respectively) of them were from politically pliable categories. The trend is a deviation from what Sarkaria Commission had expressly stated should be followed.

Further, Table 1.5 amplifies the reality that for the political appointees, that is, those selected as governors on political grounds by the ruling party or the dominant party in a coalition, the position of the governor is given as a prize for political loyalty and/or for seeking a partisan role during crises of a constitutional or political nature during the functioning tenure of the incumbent. It is apparent that political expediency rules the roost. The ruling party dispensation at the center desires to have persons of similar political background/ideology as governors in the states because it is thought that:

> in a situation of the same party in power at the centre and in the state, the Governor with political loyalty to the ruling party would

Table 1.4 Profiles of Governors by Background/Profession (1950–April 2015)

Professional categories	*Percentage*
Politicians	52.3%
Civil Servants	25.6%
Judges/Advocates	9.0
Defense Officers	6.3%
Academicians	3.9%
Others*	2.9%

Source: Table compiled by the author, based on Ashok Pankaj, "Governor in Indian Federalism – II: Hiatus between Constitutional Intents and Practices," *Indian Journal of Public Administration* 63(1) 13–40 2017 (sagepub.in/home.nav DOI: 10.1177/0019556117689850) p. 18.

Note: *Others include erstwhile Rajas/head of Princely states/freedom fighters)

Table 1.5 Profiles of Governors from Political Background/Profession (1950–April 2015)

Professional categories	*Percentage*
Former MPs, MLAs	19.4%
Former Ministers at Center/States	14.5%
Former Chief Ministers	12.9%
Trade Union leaders and Others*	5.5%

Source: Table compiled by the author based on Ashok Pankaj, "Governor in Indian Federalism – II: Hiatus between Constitutional Intents and Practices," *Indian Journal of Public Administration*63(1) 13–40 2017 (sagepub.in/home.nav DOI: 10.1177/0019556117689850) p. 18.

Note: *indicates that some of them later became president/vice-president of India

> save the state government from any political arising out either of internal rebellion or from the threat of the opposition parties. Both the phenomena increased post 1967. In the second situation of having different political parties in power at the centre and in the state, the ruling party at the centre expects a politically loyal Governor to turn any political crisis in the state to its advantage. In both situations, the Governor could play an important role.
>
> (Pankaj 2017)

The most perplexing is the practice of choosing ruling chief ministers and serving cabinet ministers as governors. As A. Pankaj states this is done quite frequently:

> The most disturbing is picking up incumbent chief ministers and cabinet ministers as Governors. Some of the recent examples are alarming. Sushil Kumar Shinde, the chief minister of Maharashtra, was first shunted out from the state and accommodated as Governor and while he was serving as Governor, he was picked up for the position of union home minister. Similarly, Shivraj Patil, serving as union home minister, was shunted out from the Home Ministry to the Raj Bhavan of Punjab. Sheela Dixit, the three-term consecutive chief minister of the National Capital Territory Government of Delhi was chief ministerial candidate during 2013 Assembly Elections that was lost by the Congress to the newly formed Aam Aadmi Party (AAP). Immediately after that she was appointed as Governor of Kerala. K. Rosaiah became chief minister of the erstwhile joint Andhra Pradesh after the sudden death of Rajasekhara Reddy. While he was appointed the chief minister, a faction within the Party was opposed to it and pressurising the party "High Command" to remove him. Impressed by the faction, he was removed and Kiran Jamadar Reddy

> was appointed as chief minister. However, Rosaiah was accommodated as Governor of the adjoining state of Tamil Nadu.
>
> (ibid., 20–23)

It is clear from the data presented here that whichever party was at the center obviously was oblivious of the Constituent Assembly's philosophy and intent on this matter (as pointed out by the Sarkaria Commission), especially since 1967.

Judicial interpretation of Article 356

The nature of pronouncements of the judiciary on the article has varied over the years in the context of cases concerning Article 356. Three approaches can be identified:

A Initially, the judiciary distanced itself from Article 356 because it categorized it as a political question that it presumed was to be dealt with by the president and his or her office.

B Then the judiciary started looking at the misuse of 356 – whether or not there is misuse.

C Lastly and lately, judiciary has adopted a substantive approach when 356 was misused extensively in the states.

The judiciary, especially the constitutional courts, have done yeoman service in reinterpretation of Article 356. Article 356 Clause (1) has been the subject matter of analyses at the hands of the Supreme Court in three of its decisions, namely, in *State of Rajasthan vs UOI* (1977)[5] and *S.R. Bommai vs UOI* (1994), and recently, in July 2016, the Supreme Court has further given a substantive additional reinterpretation in the case of the Arunachal Pradesh verdict by restoration of the Congress government in *Nabam Rebia and Bamang Felix vs Deputy Speaker SLP(C)* case.[6] Since the *S.R. Bommai* case is more comprehensive than the *State of Rajasthan* case, it would be sufficient to refer to *S.R. Bommai* alone

> In the S.R. Bommai case, the majority opinions are two, one was rendered by P.B. Sawant J. on behalf of himself and Kuldeep Singh J. The other was rendered by B.P. Jeevan Reddy J. for himself and S.C. Agarwal J. and with whose reasoning and conclusions S.R. Pandian J. agreed fully.

The Arunachal case also is discussed in this section as it sets important parameters in the federal construct in the ongoing struggle between the center and the states. In April 2016, the Uttarakhand High Court gave a landmark judgment ordering the revocation of Article 356 in Uttarakhand more or less in line with the previously mentioned judgments and has substantiated the "principle of responsible government" and the requirement of a "comity . . . to further the constitutional vision of democracy in the larger interests of the nation." The constitutional courts have thus time and again emphatically impressed on the necessity for governors to understand that they are not representatives who are elected by the masses.

To understand why the constitutional courts have come down heavily on the exercise of this article in recent times, we need to be clear about the provision as stated in the Constitution and how it gives scope for its misuse. At the outset it is to be noted that the principle of when Article 356 can be invoked sets out that if the president, "on the receipt of report from the Governor of a State or otherwise . . . is satisfied that the situation has arisen in which the government of the state cannot be carried on in accordance with the provisions of the constitution."[7] On the matter that the material placed should be sufficient to warrant the imposition of Article 356, Justice P.B. Sawant has opined in the *S. R. Bommai* case (1994) "that . . . not every situation arising in the state but a situation which shows that constitutional governance has become an impossibility – this alone shall entitle the President to issue a proclamation under Art 356."[8]

Further, "it was stated unequivocally in the *Bommai* case that the power conferred by article 356 is a conditioned power; it is not an absolute power to be exercised in the discretion of the President." Para 280 of the *Bommai* judgment states that:

> The power conferred by Article 356 is a conditioned power; it is not an absolute power to be exercised in the discretion of the President. The condition is the formation of satisfaction subjective, no doubt that a situation of the type contemplated by the clause has arisen. This satisfaction may be formed on the basis of the report of the Governor or on the basis of other information received by him or both. The existence of relevant material is a precondition to the formation of satisfaction. The use of the word "may" indicates not only a discretion but an obligation to consider the advisability and necessity of the action. It also involves an obligation to consider which of the several steps specified in sub-clauses (a), (b) and (c) should be taken and to what extent? The dissolution of the Legislative Assembly assuming that it is permissible is not a matter of course. It should be resorted to only when it is necessary for achieving the purposes of the Proclamation. The exercise of the power is made subject to approval of the both Houses of Parliament. Clause (3) is both a check on the power and a safeguard against abuse of power. Clause (1): Clause (1) opens with the words "if the President . . . is satisfied."[9]

It is clear that the precondition is the formation of satisfaction – subjective, no doubt – that a situation of the type contemplated by the clause has arisen. The satisfaction must be based on objective criteria manifest from the material placed. Several steps specified in subclauses (a), (b) and (c) of Article 356 (1) should be taken into consideration, but to what extent? The nature of the action should follow accordingly. The dissolution of the Legislative Assembly – assuming that it is permissible – is not a matter of course or automatic. It is a last resort. Moreover, the use of this power under Article 356 is subject to the approval of the action by both the houses of the Parliament. Thus, subclause (3) is both a check on the power and a safeguard against abuse of power.

Subclause (1) opens with the words "if the President. . . . is satisfied." There is a hint here that the satisfaction could be subjective. The satisfaction referred to in Article 356(1) really means the satisfaction of the union council of ministers, with the prime minister at its head. Justice B.P. Jeevan Reddy, in this same *S. R. Bommai* case, 1994[10] has given the opinion that since governor cannot himself take any action of the nature contemplated by Article 356(1), he reports the matter to the president, and it is for the president to fully understand based on material information before him – that situation of the nature contemplated by the article on emergency in the state has arisen.

In a landmark case in July 2016 namely *Nabam Rebia, and Bamang Felix vs Deputy Speaker*[11] the Supreme court has said the following, namely, that Article 356 is being blatantly misused when the government of the state is of a different political dispensation than the center. In the *Arunachal* case, the Supreme Court quashed the Arunachal Assembly Governor J. P. Rajkhowa's order to advance the assembly session and its eventual leading to the imposition of state emergency there and characterized it as a "thrashing given to the constitution and a spanking to governance."[12] It ran down Rajkhowa's plea that he was only trying to resuscitate the "democratic and legislative process" in the state. The court stated that a governor, unlike MLAs and ministers, is not elected by the people but only "an executive nominee whose powers flow from the aid and advice of the Cabinet," who are elected representatives. Further, it was argued that tenure of the governor is subject to the "pleasure of the President." Justice Khehar opined that therefore

> using discretionary powers to summon or dissolve Assembly sessions without the aid and advice of the Chief Minister and his Cabinet is plainly unconstitutional. . . . *The Governor is not an ombudsman for the Legislature nor the Speaker's mentor. The Governor cannot require the Speaker to discharge his functions in the manner he considers constitutionally appropriate.*[13]

Justice Khehar came down heavily on the growing involvement of the governors in matters political. He states categorically that

> what happens within the four walls of a political party is none of the governor's concern and that *The Governor must remain aloof from any disagreement, discontent or dissension, within political parties. The activities within a political party, confirming turbulence, or unrest within its ranks, are beyond the concern of the Governor.* The Governor must keep clear of any political horse-trading, and even unsavoury political manipulations.
>
> (emphasis in original)

Governor Rajkhowa had shown partiality to an "invalid breakaway group" of MLAs who were disqualified as per the terms of the Tenth Schedule of the Constitution. It was said point blank that

> there is no justification for a Governor to be disturbed about proceedings in connection with the disqualification of MLAs under the Tenth Schedule. Because the Governor has no role therein. . . . Any action taken by the Governor, based on the proceedings being carried on under the Tenth Schedule, would be a constitutional impropriety,

the court held in its 331 page order.[14]

According to Faizan M, Vice Chancellor, the NALSAR Arunachal case is related to the governor and his office, that is, how the governor is instrumental in the breakdown of the constitutional machinery and destabilizes a government elected to the assembly by the people and was carrying on business in office with its electoral mandate. In the original Constitution it was thought that governor would be directly elected by the people, but this provision was dropped and does not find place in the current Constitution. Now, a governor is more or less an agent of the central government, since he has no electoral or people's legitimacy.[15]

Both the Commissions on Centre State Relations, the Sarkaria Commission (1988) and the Punchhi Commission (2010) have underlined the constitutional role of the governor and state that "the Governor under the Constitution has no function which he can discharge by himself; no functions at all"(a view held by Dr. B. R. Ambedkar earlier in the Constituent Assembly).Although he has no functions, he has some duties to perform. The Punchhi Commission report (in para 4.2.14) observes that in a very limited way the governor can have some discretionary functions under Article 163(1). However, it is stated that the area of this discretion is limited. It is clear from this that the discretionary power of the governor has to be exercised with the aid and advice of the council of ministers, with no role of volition in it.

Paragraph 4.2.15 of the Punchhi Report states unequivocally that the governor has no powers under Article 163(2). It states unambiguously that "Article 163(2) gives an impression that the governor has a wide, undefined area of discretionary powers even outside situations an impression needs to be dispelled."

> It is an accepted principle that in a parliamentary democracy with a responsible form of government, the powers of the Governor as Constitutional or formal head of the State should not be enlarged at the cost of the real executive, viz. the Council of Ministers. The scope of discretionary powers has to be strictly construed, effectively dispelling the apprehension, if any, that the area for the exercise of discretion covers all or any of the functions to be exercised by the Governor under the Constitution. In other words Article 163 does not give the Governor a general discretionary power to act against or without the advice of his Council of Ministers. The area for the exercise of his discretion is limited. Even this limited area, his choice of action should not be arbitrary or fanciful. It must be a choice dictated by reason, actuated by good faith and tempered by caution.[16]

Discretionary powers, if any, that had been identified by the Punchhi Commission have been reiterated in the *Nabam Rebia* case (Arunachal Pradesh) and have specified and pinned down the role of the governor in very specific areas. The exact extract from Paragraph No. 140 (page 95) of this case is as follows:

> The Governor's discretionary powers are the following: to give assent or withhold or refer a Bill for Presidential assent under Article 200; the appointment of the Chief Minister under Article 164; dismissal of a Government which has lost confidence but refuses to quit, since the Chief Minister holds office during the pleasure of the Governor; dissolution of the House under Article 174; Governor's report under Article 356; Governor's responsibility for certain regions under Article371-A, 371-C, 371-E, 371-H etc. . . . We are of the considered view, that the inferences drawn in the Justice M.M. Punchhi Commission report extracted herein above, are in consonance with the scheme of the functions and powers assigned to the Governor, with reference to the executive and legislative functioning of the State, and more particularly with reference to the interpretation of Article 163. We endorse and adopt the same, as a correct expression of the constitutional interpretation, with reference to the issue under consideration.[17]

Thus, to classify the discretionary powers of the governor, he may do the following:

a Cause dismissal of a state government if it has lost majority
b Dissolve the state legislature under Article 174

c Give a "report of the governor" under Article 356
d Has "special functions and responsibilities regarding certain scheduled areas/states especially of the North East under Articles 371 A, 371 C,371 E, 371 H"
e Either "give his assent or hold/refer a bill for the Presidential assent" under Article 200 or
f cause the appointment of the chief minister in normal circumstances or in a coalitional context under Article 164

Thus, in the court judgment in the *Arunachal Pradesh* case, the learned judges have emphasized that the inferences drawn by the previously mentioned reports – namely, the Sarkaria and the Punchhi reports – were correct regarding the schema and functions allotted to the governor with regard to the legislative and executive functioning in the states.

Paragraph 142 of this judgment iterates that

> We are therefore of the considered view, that in so far as the exercise of discretionary powers vested with the Governor is concerned, the same is limited to situations, wherein a constitutional provision expressly so provides, that the Governor should act in his own discretion. Additionally, a Governor can exercise his functions in his own discretion, in situations where an interpretation of the concerned constitutional provision, could not be construed otherwise. We therefore hereby reject the contention advanced on behalf of the respondents, that the Governor has the freedom to determine when and in which situation, he should take a decision in his own discretion, without the aid and advice of the Chief Minister and his Council of Ministers. We accordingly, also turn down the contention, that whenever the Governor in the discharge of his functions, takes a decision in his own discretion, the same would be final and binding, and beyond the purview of judicial review. We are of the view, that finality expressed in Article 163(2) would apply to functions exercised by the Governor in his own discretion, as are permissible within the framework of Article 163(1), and additionally, in situations where the clear intent underlying a constitutional provision, so requires i.e., where the exercise of such power on the aid and advice, would run contrary to the constitutional scheme, or would be contradictory in terms.[18]

The learned judges thus summarized the position on governor's powers as follows:-

> Firstly, the measure of discretionary power of the Governor, is limited to the scope postulated therefore, under Article 163(1).

Secondly, under Article 163(1) the discretionary power of the Governor extends to situations, wherein a constitutional provision expressly requires the Governor to act in his own discretion.

Thirdly, the Governor can additionally discharge functions in his own discretion, where such intent emerges from a legitimate interpretation of the concerned provision, and the same cannot be construed otherwise.

Fourthly, in situations where this Court has declared, that the Governor should exercise the particular function at his own and without any aid or advice, because of the impermissibility of the other alternative, by reason of conflict of interest.

Fifthly, any exercise beyond the Governor's jurisdictional authority, would be inappropriate and will attract judicial review.

Another case of a similar nature regarding the misuse of Article 356, occurred in Uttarakhand around the same time and yet another instance of President's rule was imposed on 27 March 2016, leading to the fall of Harish Rawat's government. The role of both the governor and Speaker of the Vidhan Sabha in this matter came under the scanner.

The case came before the Uttarakhand High Court, which declared President's rule unconstitutional and reinstated Rawat's government back in power. The Supreme Court later conducted a floor test on the floor of the legislature in May, which led to the coming of Rawat's government back in power. Chief Justice K. M. Joseph and Justice V. K. Bist of the Uttarakhand High Court almost reiterated the same points (as mentioned earlier) that were laid down in the *Arunachal* case by the Supreme Court as cardinal guiding points to be considered by the center and the state governor before imposing Article 356 in any state. In addition, it stated very categorically that the "President is not King. The President can be an excellent person but he can be terribly wrong."[19] It expressed in addition the following points, namely:

a For the first time in legal history, misuse of Article 356 by both the constitutional and legislative entities was established and the authority of the governor and speaker in the case of Uttarakhand both were found to be against the letter and spirit of the article.

b It was held that it should not ordinarily be construed that a solitary instance be considered tangible enough to impose Article 356.

c If "corruption were to be taken into account, hardly any government could complete its 5 year term in India."[20]

The latest example of thwarted coalition government formation in Maharashtra (Nov-2019) is another case in point. Maharashtra becomes the first case in India where Article 356 was imposed because of supposed inability of political parties to form a government after assembly elections.

In Maharashtra, Governor Bhagat Singh Koshiyari did not allow the NCP-Congress-Shiv Sena coalition to come to power and instead stealthily in the dead of the night facilitated the appointment and oath taking of Devendra Fadnavis of the BJP as chief minister, who subsequently lost his position because the party did not have legislative majority. The constitution gives the governor certain freedom to use discretion to recommend the use of Article 356 in the case of a hung assembly, but before he can do that he has to explore all possibilities of government formation and has to give every party/coalition an opportunity. Koshiyari did not do so accordingly. The Congress Party and NCP both alleged that Koshiyari did not give them a chance to stake their claim to power but hurriedly installed the BJP to power surreptitiously. A constitutional jurist, P.D.T. Achary, has been quoted in *The Wire* that the governor recommended the use of Article 356:

> without exploring all possibilities. This is something that may be challenged in the courts. If there are extraneous considerations (behind the governor's move), the courts can go into those aspects. The basic test will be whether the material available with the President is enough for an ordinary person to conclude that imposition of Article 356 was inevitable. . . . This is something that the Supreme Court has to consider. The governor should have discussed the possibilities with the NCP before recommending President's rule in the state. The governor is a constitutional authority. He is supposed to exercise his discretionary powers in a fair manner. It is not a private exercise. He should exhaust all possibilities of government formations before recommending Centre's rule in a state.[21]

Our survey of the recent path-breaking judicial cases germane to our discussion on Article 356 herein states that the final interpreter of the Constitution, namely the judiciary, has held categorically that the governor is not chosen through elections, cannot be equated as a representative of the people and cannot impose constitutional provisions and their far-reaching implications based on his volition alone. Therefore, the constitutional courts have now laid a clear set of express rules (dos and don'ts) that should guide both the governor and the center in the eventuality of use or imposition of Article 356. The Arunachal and Uttarakhand examples are very important because they have reinstated opposition led governments (viz, the Congress party) in the two states which were being targeted by BJP quickly, within three months after their displacement (which also is a record of sorts for judicial pushback of imposition of Article 356 by the executive in the country). These defeats of the ruling dispensation in the cases cited above has reaffirmed the supremacy of India's Constitution and that of the higher courts, which have consistently underlined the importance of democratically elected governments in the states when there is a different political dispensation at the center.

Upshot

The preceding discussion has clearly posited the factual position on contingencies where Article 356 can be used and where it cannot. This logic was first enunciated in the well-known *S. R. Bommai* judgment of 1994 and has been reiterated both in the case involving change in Uttarakhand and in Arunachal Pradesh. In the state of Uttarakhand on 10 May 2016, the Supreme Court revoked the President's rule under Article 356, which had been illegally imposed in the state since 27 March 2016 by the BJP government at the center and allowed Harish Rawat, the ex-chief minister, to resume his stint as chief minister. The Supreme Court, by allowing CM Harish Rawat to take a "vote of confidence" in the Uttarakhand assembly, has further bolstered its earlier enunciation in the *S. R. Bommai* case. It had said therein, among other things, that assessing whether a council of ministers has the requisite strength or not in the state legislature should not be the personal opinion of any individual, such as the president or governor (the Uttarakhand case added the speaker of the assembly to this category). It should be ascertainable or demonstrable on the floor of the house. When such demonstration is possible, it is not to be bypassed or substituted by the personal opinion/satisfaction of anyone. It held that "such private assessment is anathema to the democratic principle."[22]

It is to be noted that the Supreme Court had also made it clear that there could be circumstances under which the decision to impose President's rule can be made without going for an elaborate floor testing in its *Bommai* judgment. "The sole exception to this will be a situation of all-pervasive violence where the governor comes to the conclusion – and records the same in his report – that for the reasons mentioned by him, a free vote is not possible."[23] In Uttarakhand the court mandated the floor test as essential, and this took place. After this floor test, CM Harish Rawat was reinstated to power after 46 days.

It was highlighted earlier that the current ruling elite of the National Democratic Alliance–II led by BJP has been overaggressive in its misusing the provisions of the Article 356 for its political gains. After beating the Congress in the 2014 and 2019 general elections, the BJP has been in an overdrive to seek majoritarian success any which way. In states where they have not had success, they are employing other nonconstitutional means and are toppling popularly elected governments in opposition-ruled states and small states. Ever since Arunachal, the list of states affected are growing and the recent examples of Maharashtra (November 2019) and Madhya Pradesh (2020) are ample proof that this trend may not abate in the near future. In an editorial, the *Indian Express* posits,

> However the expansion must be through democratic political means. In between electoral bouts, the BJP as the party that rules the centre can ill afford to show disrespect for, or impatience with institutions that hold up the framework of a constitutional democracy. Be it in Uttaranchal, Arunachal or in Delhi . . . the BJP has been caught transgressing vital institutional lines and norms.[24]

History is witness to the fact that the use of Article 356 to throw out governments headed by opposition parties/coalitions by arbitrary use of Article 356, which was largely a legacy of the '70s and seemed to be used a little less thereafter, especially after the *S. R. Bommai* case (1994) introduced a set of stringent conditions and emphasizing the requirement of the important component of proof of physical legislative show/strength of numbers of elected MLAs of a government formed after an election.

It seems the Congress party earlier (as also in its coalition alliances in the form of UPA I and II) and now the BJP (along with its adhesions), as if in a vengeance, have again brought the use of Article 356 into overdrive. Ironically, even though as it has been talking about cooperative federalism, the BJP has plainly, as two recent judicial pronouncements have shown, been found to be misusing the provisions of the article and the constitutional functionaries such as the president and governor and the legislative functionary such as the Speaker of the Vidhan Sabha to facilitate its political aims, something for which, as we have seen (in the previous sections) in the judicial interpretations and debates in the Constituent Assembly, this provision was not made. This may have far-reaching implications for the polity and society and even beyond. Blatant politicking in the North-East and in border states could have a destabilizing effect and an invitation to enemy countries to usurp our territory. "The dismissal of a democratically elected government should be a weapon of last resort and not one aimed at political ducks and drakes" warns an editorial.[25] Certainly, there is a growing concern at the misuse of constitution and its functionaries such as the governor, and a rethink is necessary. Moreover, it is high time that the federal philosophy that underlies at the heart of Article 356 be reviewed in modern times, especially as it was retained from a colonial-era document that has long past gone.

Notes

1 See Constitution of India Part XVIII, www.constitution.org/cons/india/p18356.html (accessed on 18 November 2018).
2 Constituent Assembly Debates (CAD), Vol. IX, pp. 132–133
3 Constituent Assembly Debates (CAD), Vol. IX, pp. 148 &177.
4 This information was given by the Ministry of Home Affairs in response to an RTI query. See https://factly.in/how-many-times-presidents-rule-imposed-so-far-india/(accessed on 24 June 2018).

5 AIR 1977 SC 1361.
6 CIVIL APPEAL NOS. 6203–6204__OF 2016 SLP(C) Nos. 1259–1260 of 2016 *Nabam Rebia and Bamang Felix vs Deputy Speaker* (July 2016), supremecourtofindia.nic.in/jurisdiction.htm (accessed on 29 July 2017).
7 Art 356, Constitution of India Ministry of Law and Justice (Legislative Department) Government of India New Delhi (2015).
8 See *S.R. Bommai vs Union of India* on 11 March 1994, Indian Kanoon, https://indiankanoon.org/doc/607 (accessed on 16 March 2016).
9 Para 280 of *S.R. Bommai vs Union of India* on 11 March 1994, Indian Kanoon, https://indiankanoon.org/doc/607 (accessed on 16 March 2016), p. 134
10 See *S.R. Bommai vs Union of India* on 11 March 1994, Indian Kanoon, https://indiankanoon.org/doc/607 (accessed on 16 March 2016).
11 See CIVIL APPEAL NOS. 6203–6204__OF 2016 SLP(C) Nos. 1259–1260 of 2016Nabam Rebia, and Bamang Felix vs Deputy Speaker (Jul 2016), Indian Kanoon, http://indiankanoon.org/doc/192490620/ (accessed on 29 July 2017).
12 See CIVIL APPEAL NOS. 6203–6204 *supra.*
13 See CIVIL APPEAL NOS. 6203–6204__OF 2016 SLP(C) Nos. 1259–1260 of 2016 Nabam Rebia, and Bamang Felix vs Deputy Speaker (July 2016), Indian Kanoon, http://indiankanoon.org/doc/192490620/ (accessed on 29 July 2017). Emphasis in original.
14 See 'SC Quashes Arunachal Governor's Order, Restores Tuki Govt. to Power', *The Hindu* July 14,2016, p. 1
15 Ravish Kumar, 'Prime Time NDTV Hindi', www.ndtv.com/video/shows/prime-time/one-more-historic-judgement-of-supreme-court-423652 (accessed on 15 July 2016).
16 Para 4.2.15, M M Punchhi Committee: Report of the Commission on Centre-State Relations (March2010), p. 61, http://interstatecouncil.nic.in/wp-content/uploads/2015/06/volume2.pdf (accessed on 15 November 2017).
17 Para 140 of CIVIL APPEAL NOS. 6203–6204__OF 2016 SLP(C) Nos. 1259–1260 of 2016 *Nabam Rebia and Bamang Felix vs Deputy Speaker* judgment (July 2016), Indian Kanoon, http://indiankanoon.org/doc/192490620/ p. 95(accessed on 29 July 2017).
18 See Para No.142 CIVIL APPEAL NOS. 6203–6204__OF 2016 SLP(C) Nos. 1259–1260 of 2016 Nabam Rebia, and Bamang Felix vs Deputy Speaker, p. 96, *supra.*
19 See 'Prez is Not King: Key Points from HC's Uttarakhand Verdict', *Hindustan Times*, 22 Apr 2016, www.hindustantimes.com/india/president-is-not-king-key-points-from-hc-verdict-on-uttarakhand/story-ygpHCKH9ceeUHSauY8Z9BL.html and Writ Petition (M/S) 795 of 2016 in the High Court of Uttarakhand, Nainital, http://lobis.nic.in/ddir/uhc/KMJ/judgement/25-04-2016/KMJ21042016WPMS7952016.pdf (accessed on 16 January 2019).
20 See Writ Petition (M/S) 795 of 2016 in the High Court of Uttarakhand, Nainital http://lobis.nic.in/ddir/uhc/KMJ/judgement/25-04-2016/KMJ21042016WPMS 7952016.pdf (accessed on 16January 2019).
21 See 'President's Rule in Maharashtra: How Constitutional were Governor Koshyari's Actions?', *The Wire*, https://thewire.in/law/maharashtra-presidents-rule-koshyari (accessed on 27 March 2020).
22 See 'Uttarakhand Floor Test: Supreme Court's Bommai 2.0 Moment', *First Post*, www.firstpost.com/politics/uttarakhand-floor-test-supreme-courts-bommai-2-0-moment-2776784.html# (accessed on 2 Jul 2016).
23 See S.R. Bommai vs Union of India on 11 Mar 1994, *supra.*
24 'Raj Bhawan Dharma', Editorial, *The Indian Express*, New Delhi, 14 Jul2016, p. 14.
25 'Playing Political Ducks and Drakes', Editorial, *The Hindustan Times*, New Delhi, 14 Jul 2016, p. 16.

References

Jaffrelot, Christophe (2000), "The Rise of the Other Backward Classes in the Hindi Belt", *The Journal of Asian Studies*, Vol. 59, No. 1, pp. 86–108.

Jaffrelot, Christophe (2003), *India's Silent Revolution: The Rise of the Lower Castes in North India*. London: C Hurst & Co.

Pankaj, Ashok (2017), "Governor in Indian Federalism – II: Hiatus between Constitutional Intents and Practices", *Indian Journal of Public Administration*, Vol.63, No. 1, pp. 13–40.

Reddy, K.J. and Joy V. Joseph (2004, Mar), "Executive Discretion and Art 356 of the Constitution of India: A Comparative Critique", *Electronic Journal of Comparative Law*, Vol. 8, No. 1. www.ejcl.org/81/art81-4.html (accessed on 2 Jan 2019).

Rukmini, S., Samarth Bansal and AMP (2016, Jan 29), "President's Rule: A Story of Over 100 Chapters". www.thehindu.com/data/President%E2%80%99s-rule-a-story-of-over-100-chapters/article14179739.ece (accessed on 22 Mar 2018).

Sarma Sarkar, Ranadhir (1986), *Union-State Relations in India*, New Delhi: Institute of Constitutional and Parliamentary Studies National.

2

INDIAN FOREST SERVICE

Mandate and state governments

Rekha Saxena and Rajesh Ranjan

In the overall discourse of center–state relations and the role of All India Services (AISs), the role of the Indian Forest Service (IFS) is a neglected and unexplored area of inquiry. Perhaps since their jurisdiction (forests of all hues) does not touch the everyday lives of the overwhelming majority of the population (as the majority lives in non-forested and heavily or moderately inhabited areas), therefore, compared to the Indian Administrative Services (IAS) and Indian Police Service (IPS), the structural and behavioral tilt and biases against the IFS cadre compared to other levels of government are not much debated and discussed nor get representation in academic or journalistic writings.

The Indian Forest Service is among the three apex All India Services (AISs) group. The other two are the Indian Police Service (IAS) and the Indian Administrative Service (IAS). Indian Forest Service (IFS) was constituted in 1966 under the All India Services Act 1951. Previously, it was the Imperial Forestry Service, which was in existence from 1865 to 1935 during the British rule in India. The officers of Indian Forest Service are enrolled through a very stringent competitive examination. Afterward, they are trained by the central government for around two years. The services of Indian Forest Service are positioned under joint cadres and various state cadres, despite the fact that they have sanction to work under both the central and state governments. Like all other AISs, the basic aim and objective of IFS is to maintain throughout the country a uniform standard of professional service. They are required to keep a national perspective even while working under the control of the state government. As such, they have certain constitutional safeguards which are enshrined primarily in Article 311 of the Constitution. The main purpose and the directive of the service is the proper enforcement of the National Forest Policy (whose stated objective is to ensure stability of environment and conservation of ecological balance essential and crucial for subsistence of all life forms – human, animal and plant), the Forest Conservation Act (FCA) and the Wildlife Act (WLPA). The Indian Forest Service officers during their field postings in corresponding state cadres work for development, conservation and protection of forests and wildlife along with the goal to exceed, intensify and enlarge the livelihood prospects of communities reliant on forests in rural and tribal areas.

IFS officers are largely independent, enjoy a great degree of autonomy from the district administration and make use of administrative, judicial and financial powers in their own sphere. Each of the key positions in a state forest department are held by IFS officers – for instance, positions like Principal Chief Conservator of Forests (PCCF), Conservator of Forests (CF), Divisional Forest Officer (DFO), etc.

The Indian Forest Service Officers are also entitled for state and central deputations, like their fellow IPS and IAS officers. Deputation of Indian Forest Service officers to the central government comprises appointments in central ministries at the rank of Director, Deputy Secretary, Joint Secretary and Additional Secretary, etc., and nomination in several Public Sector Units (PSUs), Academies and Institutes at the position of Chief Vigilance Officers (CVO), Managing Directors (MD), Inspector General (IG) and Director General (DG) etc.

Indian Forest Service Officers also go on deputations to international forums, foreign governments, United Nations bodies, international organizations, INGOs/NGOs and voluntary organizations (VOs) in addition to the private sector as per the Indian Forest Service (Cadre) Rules, 1966. The Indian Forest Service cadre comes under the jurisdiction of the Ministry of Environment, Forest and Climate Change (MoEFCC).

State government conflict with the mandate of IFS officers

Like several other federal government structures, the areas of responsibility of various levels of governments are sometimes well defined and at times ill or inadequately defined, giving rise to a situation of conflict. Let us see (in subsequent paragraphs) how inadequate and ill-defined areas of responsibility of each of the layers of government sometimes result in a situation of conflict between the mandate of IFS officers and the interests of state governments.

Schedule-VII of the Constitution of India divides constitutional responsibilities under the union, state and concurrent lists. Subjects that are responsibilities of the central government are listed in central list. Subjects pertaining to national importance as well as state importance like education, employment, forests, etc., where positive interventions are expected from the center as well as states are kept in the concurrent list. And subjects where the impact is likely to remain within the confines of states and where the state governments are best and most aptly suited to intervene are in the state list. Finally, those subjects which are not part of these categories are residuary subjects in which the center alone has jurisdiction, such as the environment.

Forest is in the concurrent list. On the other hand, land is in the state list, except for certain provisions pertaining to it, which are in the union and concurrent list. (The main entry pertaining to land in the state list according to the Constitution is as follows: "Entry 18: Land, that is to say, rights in or over land, land tenures including the relation of landlord and tenant, and the collection of rents; transfer and alienation of agricultural land; land improvement and agricultural loans; colonization.")

Forestry is governed (for which the levers of policy making and executing power largely rests in the hands of IFS cadre) by the following four acts:

The Indian Forest Act 1927, which consolidates the law relating to forests, the transit of forest-produce and the duty leviable on timber and other forest-produce. It sought to merge and reserve the areas with forest cover, or significant wildlife, to control movement and transportation of forest produce, and duty leviable on timber and other forest produce. Further, it outlines the process to be charted for announcing an area to be a Reserved Forest, a Protected Forest or a Village Forest. It describes forest offenses, the acts forbidden inside a Reserved Forest, and fines taxable on breaching of the provisions of the act.

The Forest Conservation Act 1980, which was passed to help protect the country's forests. Accordingly, it sternly limits and controls the de-reservation of forests or utilize forest land for non-forest motives without the earlier consent of central government. Therefore, the act sets in place the preconditions for diverting forest land for non-forest moves.

Wild Life (Protection) Act 1972, which was enacted by the Government of India with the aim to successfully save the wildlife and control poaching, smuggling and illegal trade in wildlife and its derivatives. However, the act was altered in January 2003, under which the penalty and punishment for crime were made more rigorous. The ministry made further suggestions in the law by initiating more strict action to harden the act. The aim is to safeguard the listed endangered flora and fauna and ecologically significant secured areas.

The Scheduled Tribes and Other Traditional Forest Dwellers (Recognition of Forest Rights) Act 2006, which acknowledges the rights of forest-dwelling Scheduled Tribes and Other Traditional Forest Dwellers on the forest regions occupied by them and give a structure for maintaining the same. Nonetheless, it is a responsibility of the Ministry of Tribal Affairs as a nodal body to execute it on the ground, but in reality due to little manpower and resources at hand, it is the Forest department in all states which plays most crucial role in its implementation, non-implementation or inadequate implementation.

It is the second and third act which are the cause of much of the problem between IFS and any state government, especially the ones which have higher forest cover and a higher number of IFS officers.

The conflict is as follows: while as per the state list (entry 18), state governments have full powers to exercise control over land, which includes their rights to settle surplus/government land to any person with or without consideration (this power is an important power exercised by the state in its role as a welfare state and by the political party in power to bestow largesse on the public to retain/increase its vote bank), the Forest (Conservation) Act 1980 greatly infringes upon this power. This is because diversion of forest land is not possible without consent of the central government, as per the provision of this act (earlier, as per the Forest Act 1927, the power to notify a forest as a reserve or protected forest rested with the state government; while this has not changed, once notified, the power to de-notify now requires approval of the central government).

Any contravention to the provision of this act, such as unauthorized diversion, is punishable by simple imprisonment, which may extend to 15 days. It is to be noted that the chief secretary of a particular state has been sent to prison due to violation of this provision. This implies that even though the state government (say the chief minister) orders his chief secretary to divert the land, the chief secretary or any such official who has caused the aforesaid unauthorized diversion is punishable under this act (section 3A). One of the main tasks of IFS, on the other hand, is to prevent any unauthorized diversion.

The animosity between the state government and the IFS is clear. While the state would be interested in diversion of land, they are not allowed to do so and view the existence of an officer of Indian Forest Service as a major symbol of loss of their power. Some verdicts of the Supreme Court have additionally aggravated this. For instance, in *T.N. Godavarman vs Union of India* (2005), the court ordered that states have to take earlier consent from the center prior to deviation of any forest land for non-forestry pursuit. Likewise, in *Centre for Environmental Law, WWF vs Union of India* (2013), acceptance from the Indian Board of Wildlife was sanctioned before de-notification of any protected region by the states.

There are a plethora of cases and instances in which state governments had come in direct confrontation with the IFS officers related to forestland diversion or a wildlife act violation, even though procedures are well defined and clearly laid out. Many state governments feel that many projects related to industry, mining, hydropower and thermal or irrigation don't see the light of the day due to the activism of Indian Forest Service officers who, under the pretext of the FCA or WLPA, send unfavorable and wrong reports that lead to non-granting or a long delay in forest and wildlife clearances, thereby limiting the developmental ambition of the state government. Although it is difficult to prove the hand of Indian Forest Service officers is involved in delays or rejections in getting forest clearances, data speaks and does prove that getting environmental clearances is a big hurdle. As per *CEEW Report: State of Environment Clearances in India: Procedures, Timelines & Delays across Sectors and States* (Gupta et. al 2017), a substantially huge number of projects in nearly every sector have unresolved forest clearances (Chaturvedi et al. 2015; Dhar and Srivastava 2015). The report claims that the generic type of industry that includes a variety of industrial subcategories like paper and pulp, iron and steel, chemicals, etc., has taken a sizable hit from the exercise, where around 90 percent of the applied projects are hanging fire. The smallest share of incomplete projects is in the hydropower sector, where not more than 40 percent of the projects are anticipating forest clearance. Normally, 40–60 percent of projects are awaiting forest clearance. An exhaustive examination tells that 35 percent of projects applications registered in 2009 and 52 percent of applications of projects applied for in 2010 did not receive forest clearance till 2015. The report also states that forest clearance is a procedure where there is a huge slowdown regardless of the state. It is also to be noted that forest clearance matters were found to be most important for thermal power, mining and industrial sectors, all in the red category. The clearance is simple in the orange and green category of work. Another report by the Centre for Science and Environment (CSE) mentions that in the case of coal mining, forest clearance has been approved more easily, though there were delays. According to a CSE report published in 2014, in the course of preceding five years (2007–2012), MOEFCC has granted Environmental Clearances (ECs) permission to double coal production capacity in the country. From a federalism point of view, it is noteworthy that as mines and minerals is a center list subject, the MOEFCC does not created much problem, but because industry belongs to the state list, the IFS cadre working in the state were uncooperative. However, such a conclusion should not be drawn hastily, as there is an enormous interest/stake for state governments (in terms of royalty and other benefits) in mineral extraction, and in the case of setting up industries, the center, too, has an enormous stake. Further, we all know that some years ago MOEFCC put strong resistance to coal mining by enunciating the concept of "Go, no-go," according to which nine coalfields were classified and categorized as Category A blocks (no-go) and Category B blocks (Go). Category A blocks met the specification of having either more than 10 percent weighted forest cover or more than 30 percent gross forest cover. In the event that both these conditions are not fulfilled, then it is considered a category B Block or "Go" area. In case of Forest Rights Act 2006 implementation, too, state government comes in conflict with IFS. State governments in general, in order to reap electoral dividends, want to distribute *pattas*, but IFS gets in the way, misleadingly citing issues with the FCA, wrongful claims (e.g., citing reasons for rejections of claims on the false ground in most of the cases as "encroachment after the cut-off date of 13 December 2005," etc.) and wildlife (particularly in sanctuary and tiger reserve areas). Collective rights claims over community forest and forest resources and other commons, barring few exceptions, are not recognized despite the pressure of state governments and sometimes even of the Ministry of Tribal Affairs (MOTA) at the center.

Another irritant for state governments in IFS's mandate concerns implementation of the wildlife act. As per wildlife clearance provisions, any project pursuit falling either in the core zone (for projects of national significance and social welfare only) or the buffer zone consisting of a 10 km radius of the core zone boundary needs wildlife clearance in advance before applying for the environment clearance. There are several cases where IFS officers applied this that created irritation in state governments. On behest of an unfavorable report of an IFS officer, NGT stopped building of a 2 km wall inside the Rajaji National Park in Uttarakhand after it realized the wall would block a corridor used by elephants along the Song river. Likewise, in case of the Yettinhole Integrated Drinking Water Project, the concerned officer exhibited fear and skepticism that the said project might adversely influence the elephant corridor and harm the habitat of other wild animals, as dams will be built in eco-sensitive regions of the Western Ghats. Similarly, NMDC was questioned by the environmental impact monitoring panel to shut down mines in the Panna district of Madhya Pradesh by 2016 to rescue big cats. There are several such cases where IFS officers, in order to implement the wildlife act, have come in confrontation with state governments. In one case they objected to Mulayam Singh Yadav's Lion Safari Project in Etwah. The proactiveness of an IFS officer resulted in NGT passing an order of closing down over three dozen illegal quarries on the edges of the Sathyamanglam Tiger Reserve (STR), which for several years had been the cause of polluting stone dust and a constant din – a decision which was not easily swallowed by the state government.

As major minerals are in the central list, many times the mandate of IFS to protect forest cover comes in conflict with the central government, too. For example, recently in the current lockdown period, the central government announced initiating auction of 40 coal blocks spread across a few coal-rich states. Anticipating there will be protest and green clearance becoming a major irritant in the process of expanding coal mining, it also initiated a parallel process of diluting provisions of the Environmental Impact Assessment (EIA) in the shape of floating a new bill that allows miners to start mining and seek clearance later on. The proposed EIA bill also aimed at weakening the public hearing process where inhabitants have space of airing their concerns around possible livelihood insecurity and the degradation of the surrounding environment in the case of giving the nod to mining. However, many activists say that the weight and power of the central government is so much that in most of the cases IFS officers representing the interest of environment and forest seem to be hand in glove with the power at the center. Many activists working on mining rights issue say that in real situations, before some very big mining companies, the center and state, political and administrative executives of all departments, various regulatory institutions and tribunals just simply fall in line. Also, contrary to the central argument of this chapter –that state government, disregarding environmental concerns, wants the diversion of forestland for expansion of industrial and mining activities (and where IFS cadre gets in the way of realizing this desire) – sometimes an altogether different spectacle occurs, as is happening currently in case of coal block auctioning, where it is the state governments of Jharkhand and Chhattisgarh that seem to be standing against coal mining in the name of protecting forests, the livelihood of indigenous communities, forestland, rivers and the elephant corridor.

In an idealistic world, one may argue that since conservation of forest and maintenance of wildlife are in the interest of maintaining the local environment, which ultimately would benefit the people, hence states should see no conflict with the Forest Conservation Act and Wildlife Act. However, in the real world there is a great reluctance on the part of the state governments to reserve land for forest. This can be seen from the rising size of the Compensatory Afforestation Fund Management and Planning Authority (CAMPA) fund, which has increased to about Rs. 54,000 cr. Seen from the perspective of the state governments, the whole approach of the Government of India in conservation of forest is inequitable. While it is all right to say that at least 33 percent of the land should be forest land, North-Eastern and hill states of Himachal, Uttarakhand and J&K often complain that they are made to sacrifice much more of their land than 33 percent for maintaining this national average of 33 percent. Hence, they want compensation from the Government of India.

The officers of IFS, who are under the control of the state government, have little voice in the top-level decision making in the Government of India. The erstwhile Planning Commission created a committee under its ex-member, Sri B.K. Chaturvedi, to go into the entire gamut of this problem. The Committee recommended (2013–2014) to give at least a fixed percent of central plan assistance to these states as compensation for having a large part of their land as forestland on which they have no control. However, with the change in government, there has been no follow-up on this report. The 14th Finance Commission also attempted to solve this problem by giving some weightage (7.5 percent) to forest cover in a state. States like Uttarakhand complain that such compensation is too little.

Here a word or two is important to say about IFS officers and local governance bodies. The Forest Rights Act 2006 gives a community (if they apply and seek CFR rights) rights of management over forest. In areas having Particularly Vulnerable Tribal Groups (PVTGs like Baigas, Pahadi Korwas, Juangs, etc.) there is even provision for Habitat Rights. If these are secured and enacted, there will be transference of ownership of forest and forest resources (except timber) into the hands of the community. But here forest department under the leadership of IFS is creating all kinds of obstacles to getting this pro-poor (tribes and OTFDs-Other forest dwellers) act implemented. Illegally they have infiltrated the claim process and thus influence the process in such a manner so that their control and hegemony over forests remains intact. Utilizing the CAMPA fund, they are now engaged in reviving Joint Forest Management Committees (which a forest employee heads and calls the shots for) and in some reported cases illegally gave them CFR and management rights, which is a gross violation of the FRA. As per the act, the governance and management of the community forest is to be entrusted in the hands of Forest Management Committees, which are to be constituted under FRA.

The mandate of the IFS has acquired an advantage also due to an overall (developing) structural tilt of the Constitution and overall national and international rising concerns toward the environment and climate change. This affects and exacerbates the previously mentioned unease in relation to the cadre at both the state government level and (wrongfully) at the local governance level. We all know that although the Constitution was environmentally blind, but, politically sensitive to the adoption of federal form of governance and therefore respecting the federal principle, it assigned all the natural resources – land, water, forests and mineral resources – to the states and limited the role of the union government to only two matters related to natural resources management, viz, the regulation of mines and mineral development and the regulation and development of interstate rivers and river valleys. Air and the environment in general become residuary subjects over which the union government alone has jurisdiction. But this paradigm started shifting since the mid-1970s. The Constitution was amended to provide important space to environmental issues. The reason was a global environmental movement culminating in UN conference on the Environment in Stockholm in 1972. Second, the global agenda of action influenced the national plan of action. Third, the political leadership of India remained deeply committed to the cause of the environment and played an important role in connecting the environment with issues of sustainable development. Fourth, public opinion against reckless use of natural resources is strong. The paradigm shifts not only resulted in the incorporation of the environment into the constitution but also redefined the powers and responsibility of the union and state government on environmental matters. The 42nd Amendment brought forests and protection of wild animals from the state list to the concurrent. Further, Article 48A was inserted in Part IV (DPSP) and ART 51 A on Fundamental Duties. The second process by which environmental legislations came to occupy the statute book was through the international treaties, conventions and protocols that India signed and ratified over years. All these created obligations for developing appropriate domestic legal and institutional framework for implementation of these treaties and conventions by the union government. Environmental jurisprudence developed, too. All these, in a way, made the center powerful in relation to state governments in matters of forest, wildlife and the environment, which emboldened the IFS cadre in dealing with state governments.

For further exploration of the subject of "uneasiness of state governments with the mandate of IFS," we discussed the subject with a few middle and senior level IFS officers working in Chhattisgarh and Jharkhand (Interviews of Kanker-DFO, Godda-DFO, Kanker-CF, Ex-Forest Secretary-Jharkhand, Dy General Manager (Exploration)-CMPDI, Convenor- Environics Trust, New Delhi and Secretary General-Mines, Minerals & People, Visakhapatnam). Some of the main points which emerged out of this discussion are as follows. First, the general view among the IFS cadre is that they are most "unwanted" by the state governments as compared to other two All India Services. Second, "forestland diversion" is the main bone of contention between them and the state governments. The state government puts enormous pressure on them to write recommendations for forestland diversion for setting up industries, mining, irrigation and hydel power projects and sometimes for setting up social infrastructure. However, with the emergence of orders regarding "linear projects" (roads, high tension electric lines, railway lines, etc.) and the provision of securing 2 hectares of forestland in nonconflict districts and 5 hectares in conflict districts without much efforts for a project has smoothed out the process and reduced the tension between IFS and state governments. Third, as compared to most of the state governments, the center is more concerned and committed toward the forest, environment and wildlife.

References

Centre for Environment Law (WWF) v. Union of India and Ors 2013.

Chaturvedi, Vaibhav et al. 2015. *State of Environmental Clearances in India: Procedures, Timelines & Delays Across Sectors & States.* CEEW Report, New Delhi.

CSE Report. 2014. *Forest & Environment Clearances: Problems for Economic Growth or Problems for Environment Protection.* New Delhi.

Dhar, Chakrabarti P.G. and Nidhi Srivastava (eds). 2015. *Green Federalism: Experiences & Practices.* New Delhi: The Energy and Resources Institute and Forum of Federations.

Gupta, Vaibhav et al. 2017. *State of Environment Clearances in India: Procedures, Timelines & Delays across Sectors and States.*

Interviews of Kanker-DFO, Godda-DFO, Kanker-CF, Ex-Forest Secretary-Jharkhand, Dy General Manager (Exploration)-CMPDI, Convenor- Environics Trust, New Delhi and Secretary General-Mines, Minerals & People, Visakhapatnam.

T.N.G. Godavarman Thirumulpad vs. Union of India & Ors 2005.

3

TELECOM REGULATORY AUTHORITY OF INDIA

Institutional Autonomy, Net Neutrality, and Cyberspace Governance

Niraj Kumar and Mahendra Prasad Singh[1]

This chapter explores a relatively understudied topic of the telecommunication sector in India from the perspective of political science and public administration, which has undergone rapid and rather radical transformation since the onset of comprehensive economic liberalization in 1991, when the state-led industrial sector in India, riddled with inefficiency and corrupt neo-feudal rent-seeking, was forced to open up to market forces, national and multinational. Telecommunication is a union or federal jurisdiction under the Constitution and was a government monopoly until 1991. The central aim of this chapter is to bring under the analytical scanner the supposed transition of the process of *government* to *governance* in this policy domain and the attendant challenges and opportunities. The transition from direct bureaucratic control to an indirect regulatory mechanism is supposed to improve transparency, predictability, accountability and autonomy of the regulator. In India, this transition is, however, far from completed yet.

The problem, questions and hypotheses

The central question raised and sought to be answered in this research is whether the transition and transformation propositioned above have actually happened in India. Is the emergent process of governance transparent, participatory and predictable? Is the process and practice of appointment of members of the regulatory authority conducive to its autonomy? Has it tilted in favor of the government and its favored corporations – public or private – rather than ensuring a level playing field for all? What are the causes of poor quality of telecom services? Are telecom administration and services being provided independently and impartially, and can the state of affairs be reformed? Our findings are mixed, with positive and negative points, largely verging on the unsatisfactory.

The Telecom Regulatory Authority of India (TRAI) was set up under an act of Parliament in 1997. It was granted regulatory powers and policy making for some matters, while licensing continued to be vested in the government's Department of Telecommunication (DoT). Thus, India has adopted the ministerial-bureaucratic process for policy formulation and implementation. Its regulatory process is weakened by lack of transparency and autonomy as well as a multiplicity of regulatory units such as DoT, the Telecom Commission predating TRAI, TRAI, cabinet minister, Cabinet and the prime minister.

Post-liberalization, the expected sufficiently competitive market has not emerged. Wireless services have been dominated by a few private companies like BSNL, MTNL and a private company, Bharti. Among internet service providers, BSNL is in the lead, followed by Reliance, MTNL and Bharti (Kaur and Neena, 2016). Some experts in a general context observe: "In a market dominated by few players and lack of effective regulation, competition is unlikely to emerge, thus increasing the likelihood of anti-competitive behaviour on the part of the dominant players" (Singh, Soni and Kathuria, n.d.: 30). The government has opened the doors for 100 percent foreign direct investment (FDI) in the telecom sector, but the sector still is largely national, with very few foreign investments flowing in. Industry observers attribute it to lack of regulatory clarity and predictability (Krishna and Machad, 2013).

Competing constitutional models

To relate our case study of TRAI to the larger debate on competing models of constitutionalism in comparative politics, there are, broadly speaking, two such models: (i) British constitutionalism *a la* Albert V. Dicey; (Dicey, 2008), and (ii) US constitutionalism *a la* Alexander Hamilton, James Madison and John Jay (Hamilton, Madison and Jay, 1987). Through the maze of common law, the parliamentary law and constitutional conventions in the United Kingdom, Dicey brought out three fundamental premises of the British Constitution, namely, parliamentary sovereignty, rule of law and constitutional conventions. In the British scheme of things, the Parliament is supreme. Neither the Crown nor the court can sit in judgment over the wisdom of the Parliament. The court can, at the most, interpret the law of the Constitution in terms of the meaning and intent of the laws made by Parliament and review administrative decisions. The situation remains unchanged in theory despite British membership in the European Union and impending exit, Scottish parliamentary devolution, and the creation of the Supreme Court outside the House of Lords since 2009. In the absence of a written basic law there is no constitutional limitation on Parliament. It is another matter that the British Parliament always respects common laws. The executive, too, is perpetually responsible to Parliament, and ultimately to the people. All in all, the British constitutional principle is essentially predicated on parliamentary majoritarianism, at least in principle. In practice, the Parliament has been exceptionally obeisant to the rule of law, conventions of the Constitution and the common law.

US constitutionalism, on the other hand, detests the "tyranny of the majority" and can be best described as consensual federal democracy. The three leading makers of the American Constitution who authored *The Federalist* borrowed the theory of *horizontal* separation of powers between the executive, the legislature, and the judiciary from Montesquieu and John Locke and created their own theory of *vertical* division of powers between the federal and the state governments. Separation of powers and division of powers combined to create the system of checks and balances that is the hallmark of American constitutionalism.

In both the foregoing models there was little scope initially for independent regulatory authorities in concept and theory as sites of independent rule making, the executive, and executive powers and functions. In the United States the idea was resisted due to primary locus of these powers in the cardinal governmental organs of the Congress, the presidency, and the Supreme Court. In the United Kingdom, too, the idea initially was an anathema due to the British obsession with the doctrine of parliamentary sovereignty. However, in course of the evolution of the process of government on both sides of the Atlantic, independent regulatory agencies gained gradual functional relevance and even came to be grudgingly granted considerable autonomy and authority.

Where does the Indian Constitution stand in this regard? It is our argument that the Indian Constitution holds the middle ground between the foregoing two models. In the Constituent Assembly of India there were strong advocates of parliamentary sovereignty in H. V. Kamath, R. N. Singh and P. S. Deshmukh, and not least Jawaharlal Nehru. Not that they were unaware of the implications of fundamental rights and federal division of powers between the union and states that entail judicial review of parliamentary enactments and executive orders, which obviously spell some limitations on the supremacy of the Parliament (India (Republic), 2003a: 1644–1666). Yet they seemed to be enamored of the idea of parliamentary supremacy, especially in public policy making and the all too important matter of constitutional amendments. Ambedkar also implied that the Indian Constitution was, in a way, two in one: federal in normal conditions and unitary in constitutional emergencies, when the Union Parliament substitutes the state legislatures (India (Republic) 2003b: 31–44). Rajendra Prasad, if at all, in his final presidential remarks in the Constituent Assembly showed a somewhat greater understanding and appreciation of the autonomous agencies created under the federal constitution (India (Republic) 2003c: 984–995). The text of the Constitution in its actual provisions clearly reflects these understandings of the leading lights of the Constituent Assembly. Nonetheless, the

Constitution itself created some such independent regulatory authorities in the Auditor and Comptroller General of India, the union and state Public Service Commissions, the Finance Commission, and the Election Commission of India.

TRAI: composition and functions

The Telecom Regulatory Authority of India (TRAI) comprises a chairperson and no more than two full-time members and no more than two part-time members. All are appointed for a term of three years by the union executive, avoiding persons having any conflict of interest, financial or otherwise, that may prejudice their function. Members, including the chairperson, may be removed for insolvency, conviction for moral turpitude, physical or mental incapacity, developing reasons for conflict of interest and abusing power and position. The TRAI is empowered to make recommendations to the government regarding the working of the sector either *suo motu* or on request from the government. Its recommendations under section 11(a) of the act are not binding on the government, but on service providers they are.

The union executive is also required under the provisions of the 2000 Amendment to the 1997 Act to set up an Appellate Tribunal consisting of a chairperson and not more than two members for a term of three years in consultation with the Chief Justice of India for dispute settlement on an application made by the "Central Government or a state government or a local authority or any person" "aggrieved by any direction, decision, or order made by the Authority" (Section 14A of the Act). The arbitration of the Appellate Tribunal is subject to appeal to the Supreme Court. The TRAI itself is obliged to make an annual report to Parliament. Thus, the adjudicatory function is now separated from the regulatory function.

Regulatory autonomy

The various regulatory reforms have come through a good deal of trial and error and experimentation. In some cases regulators were appointed because the government and Parliament thought – in some cases on the basis of their own experience – that ministries and their bureaucracies could not be objective enough in dismantling a government monopoly and bringing in private players fairly, and in some cases courts intervened and ordered autonomous regulatory regimes because they thought that autonomy was greatly whittled down, despite claims to the contrary. However, in a sector like civil aviation, the ministry has proceeded with reforms under its own control. Private participation in the previously state sector-dominated economy has brought enormously large investment to financially starved public sectors.

However, the assessment of the regulatory regime has been mixed, often contradictory. Pradip Baijal, a former chairman of TRAI, gives a rosy picture, a state of affairs also evident in the annual reports of the regulatory agency. Eighty percent expansion of the sector in just 10 years since the reforms is juxtaposed against mere 1.9 percent in the previous 50 years (1948–1998) (Baijal, 2007). However, Ashok V. Desai, in a pioneering systematic study on the role of the regulatory regime in the telecom sector, has a more critical and reconstructive evaluation. He concludes,

> The most striking feature of the regulatory system is the relationship between the regulator and the government. The two had considerable friction in early years. The friction went down in later years, and the two worked remarkably in tandem in 2001–03. But in the process, the regulator became very mindful of the government's interests. It was not a case of complete regulatory capture by the government; but in the case of the access deficit charge and the USF [Universal Service Fund], the regulator implemented the government's intentions.
>
> (Desai, 2006d: 135–136)

The regulatory regime in the telecom sector was the third to be set up, first by union executive decision and finally by an Act of Parliament, after such bodies as the Security and Exchange Board of India (SEBI) in 1988 and Insurance Regulatory and Development Authority (IRDA) post-1991. The Central Electricity Regulatory Commission was put in place in 1998. More are being contemplated. In the case of TRAI, commissioners have been either serving or retired bureaucrats or fellows or directors of research institutes with close ties with the parties in power. The chairman, too, is recruited from the same catchment area and with the same background.

The TRAI is mandated with four basic functions, among others, under the parliamentary statute establishing it:

- Ensuring interconnectivity between competing telephone operators with the choice of subscribers to select any.
- Ensure free competition and prevent monopolistic tendencies by practices such as predatory pricing, etc.
- Dispute settlement among the competing operators.
- Fix and administer uniform service obligations, by and large.

(TRAI Act, 1997)

Partial privatization of the telecom sector showed immediate good results. A mere 2 percent annual growth in the sector in the preceding half a century rose to 6 percent in just one year – 2006–07. In the assessment of one observer at close quarters, the impact was spectacular and sustained:

> Aggressive action by the newly appointed regulator in 1997 and actions of future regulators, sometimes supported by the government, changed all that with both public and private players in this network becoming very aggressive competitors for overall benefit of the country, its economy and particularly for the consumer. From almost nil presence in 1998, and 15 per cent in 2003, private sector now runs 65 to 70 per cent of its network and their total investments during the last decade are now gradually approaching the investments made by the public sector incumbents during the last 60 years.
>
> (Baijal, 2007: 8)

This observer goes on to point out that the trend of reforms also shows that the growth of the private sector has been supplementary rather than predatory to the public sector. If anything, competition has made the public sector more efficient, and tariffs in various segments of the telecom networks have gone down by a whopping 90 percent, the lowest in the world. The growth in the sector has also been inclusive, at least in urban India (teledensity in Delhi exceeding 90 percent), and rural areas are also not totally excluded (Baijal, 2008: 17–18).

Under the new regulatory regime, scarcity, waiting periods and high charges have considerably either disappeared or gone down. However, powerful operators like Mahanagar Telephone Nigams often bypassed the TRAI or ignored its orders on arguable grounds. The TRAI found the Department of Telecommunication in the Ministry of Communications and Information Technology in the Government of India hostile to it. The government proceeded to promulgate an ordinance in 2000 preparatory to an amendment to the concerned law. The entire membership save one was dismissed. The amendment provided for an appellate tribunal, subject to an appeal to the Supreme Court only (excluding now the High Courts, whose conflicting interpretations have emasculated the regulator) and obliged the government to consult TRAI on all licenses in deference to judicial rulings. The move was aimed at bringing about some order out of chaos and providing greater autonomy to the regulator (Desai, 2006a: 155–156). Yet the semi-judicial status enjoyed by regulators like Central Electricity Regulatory Commission (CERC) until 2003 eluded the TRAI.[2] The world's fourth largest telecom network after China, the United States and Russia deserves a better deal. TRAI has recommended from time to time comprehensive reforms on licensing, access to service providers and users' service and satisfaction. It should have a greater autonomy from political interference.

It is difficult not to agree with the diagnosis of Ashok V. Desai that the restrictively narrow power granted by the government to the TRAI has "led on the one hand to encroachment by judicial bodies on the regulator's sphere of operation and on the other to making the regulator complicit in government's policies, both good and bad." Desai surmises that despite the expansion of the sector, the "weak and fractured regulatory regime could not carry out its mandate to prevent concentration of ownership of the corporate houses that got 17 licenses in 1996, six ended up holding 66 out of 84 licenses in 2002, and concentration continued unabated since that year" (Desai, 2006b:141). Desai's passing observation on the regulatory regime in general is that Indian regulatory authorities by and large have been ineffective due to the obsolete role of government persisting, weak design of the authorities created and multiplicity of agencies. To quote Desai at some length:

> The government has tendency to set up too many regulators without thinking about industry boundaries. In the financial market, for instance, there are already five regulators: Reserve Bank of India for banks, Department of Company Affairs for Companies, Securities and Exchange Board of India for stock markets, Insurance Regulatory and Development Authority for insurance, and Forward Markets Commission for non-financial forward markets. The proliferation of regulators creates problems of coordination and turf.
> (Desai, 2006c:158)

Desai's proffered suggestion is the creation of more autonomous regulatory regimes and opening up of industry to local competition by de-licensing "last-mile operations." Desai does not, however, pause to even cursorily examine the consequences of a unified and overloaded regulatory regime nor, for that matter, offer an alternative scheme of a smaller number of authorities nationally apportioning regulatory spheres or roles.

Moreover, there is the market and global side of the problem that Desai also seems to overlook. The recent global credit crisis has demonstrated again that our regulatory structures are not properly aligned to the imperatives and opportunities of the global financial market. D. N. Ghosh concurs with Desai on the point of "split regulation" but directs attention to the phenomenon of market-captivated regulators as against Desai's complaint against state-captivated regulators. In order to deal with the adverse effects of global financial crises, Ghosh recommends that the central bankers devise a unified regulatory regime that "does not shy away from interventionist regulation" (Ghosh, 2007: 4094–4095).

The Second Administrative Reforms Commission (India (Republic), April 2009) has also examined the existing regulatory regime set up since the early 1990s and underlined the need for creating an effective regulatory framework. It suggests greater uniformity in the terms of appointment, tenure, and removal, as also the degree of autonomy among the various regulatory authorities. The chairmen and members for all such authorities should be empaneled by a selection committee for appointment by the government. The composition of the selection committee should be provided for in the respective acts on the patterns of the Central Electricity Regulatory Commission Act, 1998. Parliamentary accountability of the regulatory regime should be ensured through the respective departmental parliamentary standing committee (India (Republic), ARC-II, 2009: ch. 6).

It is necessary to add here that the new institutional experimentation with the emergent regulatory regime cannot be oblivious of the principles of representative democracy and parliamentary–federal governance under the Constitution of India. The claims of political democracy and capitalist development must be reconciled. It is a reflection of the times when the state has failed and the imperatives of the market economy are readily advocated and accepted in the contemporary discourse in the Indian political economy. If uncritical statism has landed us in the developmental wasteland, market fundamentalism in due course may lead to greater distortions of the concepts of rights and justice enshrined in Parts III and IV of the Constitution. Deep down, our constitutionalism is wary of the breakdown of both political and economic orders. This is evident in the emergency provision on the national, state and financial fronts (Articles 352, 356, and 360). The ultimate accountability of the regulators to Parliament under judicial and public scrutiny cannot be wished away. Its absence would amount to whittling down democracy vis-a-vis capitalism. If the excesses of populist democracy via its distortions has proved to be the Waterloo of the state-led economy in some decades past, market fundamentalism can be a greater disaster in the long run if democratic and developmental imperatives are not critically and constructively harmonized, both at the levels of the nation and federation and "glocalization." (globalization plus localization).[3]

Under the 1997 act, both the regulatory and adjudicatory functions were fused in the TRAI itself. The 2000 amendment to the act set up the Telecom Dispute Settlement and Appellate Tribunal (TDSAT), a sector-specific specialized body for the purpose. In the comparative literature on autonomous regulatory agencies, there are, broadly speaking, three main methods of dispute resolution in the telecommunication sector; two of these are relatively more formal, and the other is informal. The formal methods include court-based adjudication and regulatory-based adjudication, and the informal one consist of arbitration, mediation and negotiation. The Indian system is, by and large, based on regulatory-based adjudication (Prasad, 2011a: ch. 3). An appeal against the award of the TDSAT may be made within ninety days in the Supreme Court of India only when a point of law is involved. The amended act provides for a tribunal consisting of a chairperson and two members. The chairperson must be a serving or retired judge of the Supreme Court or a chief justice of a High Court. The two members must have been either a secretary to the Government of India for at least two years or experts in the fields of technology, the telecommunications industry, commerce or administration. The tribunal decides by majority, and its deliberations are treated as judicial proceedings and orders as decrees of a civil court.

A study of the structure and working of the TDSAT has found that it is sound inasmuch as it is a collegial body with security of tenure and insulation from "the whims and fancy of political masters." It goes on to say: "In the matter of funding of its activities, however, this body still has to look up to the government, which certainly is a drag to an extent on its independence" (Prasad, 2011b: 74–75).

The aforementioned study concludes that the "telecom statute in India does not provide any preeminent role to the Minister in matters falling within the jurisdiction of the regulator or the telecom tribunal" (Prasad, 2011c: 76). Yet it is difficult not to feel that the regulator is less fortified against political interference and excess. It is amply demonstrated by the 2-G spectrum allocation scam that surfaced toward the end of the year 2010.

The report of the Comptroller and Auditor General of India (CAG) has revealed that the Department of Telecommunications (DOT) under the Telecom Minister (Dravida Munnetra Kazhagam/DMK) in the Congress-led United Progressive Alliance government allotted 2-G spectrum licenses to certain private companies at throwaway prices, disregarding the government allocation rules, ignoring the advice of the prime minister, ministries of law and finance, as also the recommendations of the TRAI. These irregularities cost the national exchequer a presumptive loss of Rs. 1.76 lakh crores, making the 2-G scam presumably the biggest case of political corruption in independent India. The CAG report rued: "The role of TRAI would also appear to have been reduced to that of a hapless spectator as its recommendations were either ignored or applied selectively" (Mahaprashasta, 2010: 19).

The 2-G spectrum allocation scam turned into a major row between the government and the opposition that practically paralyzed the winter session of parliament in 2010. The opposition demanded a probe or enquiry by a Joint Parliamentary Committee (JPC), whereas the government argued that the Public Accounts Committee (PAC) of Parliament was already seized with the problem. The prime minister offered to appear before the PAC, even though he is not required to do so under the rules, which in case of a JPC is mandatory. Repeated negotiations led to no consensus on the matter. When the opposition appeared determined to continue the agitation in Parliament and in the streets to the extent of forcing a mid-term election on the government, the latter blinked and offered to convene a special session of Parliament to discuss the matter and threw hints that it might be prepared to set up a JPC as well. The JPC was finally formed. On the other hand, a public interest litigation in the Supreme Court resulted in the order of the court to the Central Bureau of Investigation (CBI) to begin inquiry into the case and report directly to it over the head of the government. In the meantime, the PAC also completed its probe but submitted a fractured report by a last-minute realignment among parties manipulated by the Congress-led ruling coalition. By the autumn of 2013, the JPC also submitted a fractured report, with a majority siding with the government but as many as seven strong notes of dissent by opposition parties. However, the Supreme Court in a public interest litigation had already ruled (2 February 2012) the government's 2G spectrum allocations to the private companies "unconstitutional and arbitrary" and quashed them all. Legal proceedings against the jailed telecom minister, A. Raja (DMK), in the Congress-led United Progressive Alliance government at the center, some other party MPs and bureaucrats in a CBI court gave the individual accused the benefit of the doubt in December 2017 for the failure of the CBI to produce clinching evidence as to their crime. The CBI said it would go in appeal to the High Court.

The avalanche of scams such as those relating to the Commonwealth Games, the Adarsh Housing Society (Mumbai) meant for war widows but cornered by the high and mighty in the army, civil services and politics and a number of cases relating to judges, etc., besides the 2-G scam, made 2010 the year of exceptionally endemic corruption in the government and other high places. These developments, including the Nira Radia Tapes (a corporate lobbyist successfully seeking the allocation of telecom ministry to A. Raja of DMK) draw attention to the growing crisis of democracy in India in transition from the License-Permit-Quota Raj to crony capitalism. Unless democratic institutions, civil society and social movements intervene, democracy in India appears to be in peril.

From the foregoing controversial and scam-marred past, the TRAI has tried to evolve a more defensible procedure of policy-making and implementation in appropriate relationship with the DoT. The hallmark of this procedure is that the TRAI issues a consultation paper seeking the comments of the stakeholders and then holds an Open-House Discussion with them before firming up its recommendations to the government. To sample it, one may refer to the issue, for example, of the TRAI's recommendations on Reserve Price and associated conditions for auction of 700 MHz, 800 MHz, 900 MHz, 1,800 MHz, 2,100 MHz, 2,300 MHz and 2,500 MHz bands of spectrum as narrated, among others, in Part II of the Annual Report of TRAI for 2015–2016 placed in the Parliament. TRAI also recommended that all allocated spectrums, both commercial and spectrum allocation to various PSUs and government organizations, should be done by an independent agency. The recommendation has a federal dimension as well, which is evident from the reserve prices for various spectrum bands recommended that vary from state to state, as tabular data given in the Report show (Telecom Regulatory Authority of India, 2015–16).

However, studies continue to voice critique of both the structural parameters of the regulatory regime in India and the actual practice of the agency as well as the government's dealing with it. A survey of stakeholders' perceptions reported that 38 percent of its sample felt that TRAI needed greater functional autonomy, 24 percent were for increased financial autonomy and

38 percent were for enhanced autonomy for its appointment process. There is at present palpable governmental interference. Even though its funds are drawn from the Consolidated Fund of India, it is unnecessarily routed through DoT. Besides the first chairman, Justice S.S. Sondhi (Retd.), only government officials of the Indian Administrative Service (IAS) have been appointed by the executive, without the involvement of the relevant parliamentary committee or its chairman (CPR South Young Scholars, n.d.).

The head of Cyber Initiative at the Observer Research Foundation, New Delhi, has observed:

> TRAI changed hands during the United Progressive Alliance's (UPA) ten-year term, from one bureaucrat to another. Two of them even served as secretaries in the Department of Telecommunication, their appointments reflecting the UPA's brazen attempt to align TRAI's views with that of the government. Over the last decade, TRAI has become a veritable arm of the government, its façade of autonomy masking the political projects that the regulator has been drawn into. (Sukumar, 2015)

The same pattern continued under the Bharatiya Janata Party-led National Democratic Alliance (NDA) government, which appointed R.S. Sharma, Secretary, Department of Electronics and Information Technology, as the chairman of TRAI. In fact, the first act of the Narendra Modi government, after it was voted to power in 2014, was to amend the TRAI Act to abolish the two-year cooling off period for the chairperson to accept another appointment under the government in order to appoint Nripendra Misra as Principal Secretary to the Prime Minister (Sukumar, 2015).

Net neutrality

Since the incremental technological invention of the internet in the United States in 1969 and its expansion on a global scale since then, the principle of what has come to be called "net neutrality" has developed, which stipulates that governments should mandate internet service providers (ISPs) to treat all data on the internet on the same footing and not discriminate or charge differently by user, content, website, platform application, type of attached equipment or method of communication. The general trend initially evolved in favor of net neutrality, barring exceptions like China, where the notion of an open internet has been effectively killed by the vest censorship apparatus of the state. Elsewhere, when a user pays an ISP for an internet plan, one has the right to access all online contents – videos, games, news, social media sites – at the same broadband speed one has signed up for. In 2005, a small phone company, Madison River, based in North Carolina, began blocking its subscribers making phone calls using the internet application Vonge, its competitor in the phone call market. Consumers complained, and the Federal Communications Commission (FCC) promptly fined and stopped this anti-competitive move. In 2007, the US giant Comcast Corporation, in Philadelphia, was reported to be delaying the upload of files of Bit Torrent Inc., in San Francisco, in its peer-to-peer protocol/service. FCC intervened in August 2008 to disallow this discriminatory practice. The issue stirred up a furor in 2015, and President Barack Obama urged the FCC to uphold net neutrality. The FCC strongly ruled in prohibiting any ISP from blocking, throttling or giving any preferential treatment in terms of speed to a content provider who has paid more than another.

The foregoing net neutrality rule has changed under President Donald J. Trump, with the FCC voting against it on December 14, 2017. The demands of its reversal by net neutrality activists, including US scientists who invented the internet and by the public, have been firmly ignored. The protesters allege that the new rule is harmful to both digital democracy and competitive corporate capitalism, as also to entrepreneurial and technological inventiveness. It is also being strongly challenged by sections of the Congress and state governments. The campaign is on for its congressional repeal and/or challenge in the courts. Canada and certain European Union countries also generally follow the principle of net neutrality. The European Union introduced rules in 2015 to ensure that all contents are treated equally and there are no fast or slow lanes, except for some specific types of contents such as critical health care data.

In India, the TRAI issued a consultation paper on this issue in March 2015 that did not reach any conclusion. With a release of a comprehensive consultation paper in 2017, it is now making headway in the net neutrality debate, exploring how to define core principles, regulate traffic management practices and monitor for violations. It has so far seemed to favor net neutrality. Broadly speaking, TRAI has said it does not want any content discrimination. It does not want a market where the poor get some basic minimum internet services telecommunication companies allow and the rich get fuller internet that is faster. It disallowed in 2016 the "zero-rating" service, Facebook's Free Basics, a tempting limited data plan for customers that favored their own paid websites and services, arguably disadvantaging smaller, less popular websites. Critics say the "zero-rating" schemes give ISPs an opportunity to create walled gardens that favor their own affiliated content; it negatively affects user privacy and security as ISPs use deep packet inspections to determine if issues are on zero-rated sites and disallow the use of encryption, as it is anti-competitive and anti-innovation, favoring established sites and services that can pay to have contents zero-rated. TRAI chairman R. S. Sharma told Pranav Mukul in an interview:

> If someone misuses CDN [content delivery network] to violate net neutrality then we will figure out. All the exemptions and exclusions are meant to ensure faithful implementation of the net neutrality

> principle, and will have checks and balances. In furtherance to that, if someone tries to shortcut or defeat the spirit of net neutrality it is mischievous and accountable. If you can't do something directly, you can't do it indirectly as well, that is the basic principle.
>
> (The Indian Express, 2017)

The TRAI has mandated that ISPs will refrain from deploying any discriminatory practices like blocking, degrading or slowing down certain web traffic while proffering any specific content or data. The rule was crystallized after a series of consultations with stakeholders between 30 May 2016 and 4 January 2017 (livemint, n.d.).

The Government of India's strong support for net neutrality is evident in the action of the Competition Commission of India (CCI) imposing a penalty of Rs. 136 crore on Google in January 2018 for abusing its dominant position in online general web search advertising services in India on a complaint filed by Matrimony.com and Consumer Unity and Trust Society in 2012. The CCI found appropriate the imposition of "penalty at the rate of 5% of their average total revenue generated from their India operations from different business segments for the financial years 2013, 2014, and 2016 as provided by Google during its submission" (The Times of India, 2018).

The recent events relating to data breach scandals involving the Facebook, Cambridge Analytica and the Canadian firm the AggregateIQ have brought into the limelight the serious absence of regulations and governance globally as well as nationally in the internet space. It was revealed that Cambridge Analytica had access to personal data of over 87 million Facebook users, including 5 lakh Indians. The data so breached were used to build detailed profiles, on the basis of which specifically targeted messages were sent to people to influence the last American presidential election. Mark Zuckerberg, the founder and CEO of Facebook, subsequently admitted this breach of trust in Congressional hearings and also apologized for it and promised greater care and security, though his lawyers had earlier argued that nothing was done without the consent of users. Arghya Aengupta, a legal expert and a member of the Committee of Experts appointed by the Government of India to draft a data protection law, laments: "The real culprit is a lax regulatory environment that astonishingly deems large-scale, unethical data sharing to be legal." He goes on to advocate:

> Strong rules to ensure informed consent, greater accountability for data breaches and liability for platforms for consent, might be good places to start. If India does so, it might become a model for the global South to follow. Such a model is our best hope to cope with the brave new world that the Facebook episode demonstrates, which is no longer a distant future but rather our frightening present.
>
> (Sengupta, 2018)

Toward the end of 2017, TRAI released a set of recommendations on net neutrality for internet service providers that attempted to make for an open and transparent internet in India. This stands in stark contrast to the controversial decision of the US Federal Communications Commission's (USFCC) chairman Ajit Pai to reverse the Obama-era open internet order. In the opinion of two experts:

> The recommendations successfully walk the tight rope of balancing both industry and public interests. TRAI does this by recognising and safeguarding an end users' interest to access content in a free and fair manner while being mindful of the operational necessities of service and content providers.
>
> (Mohanty and Srikumar, 2017: 1)

The TRAI in the foregoing order has recommended the creation of a multitaskholder body for the monitoring of violations of net neutrality and enforcement of rules somewhat similar to self-regulatory bodies in India like the Advisory Standard Council of India (ASCI) and the Brazilian Internet Steering Committee, which was set up statutorily to collaborate with the regulator in promoting principles of innovation and access. This newly proposed body in India will comprise internet service providers along with other stakeholders ranging from content providers to civil society and academia. The critics have complaint that this body would overly represent the government and hence be vulnerable to political hijack. It seems that the TRAI has been inspired by the success of its open-house discussions and is seeking to institutionalize this procedure through this mechanism (Mohanty and Srikumar, 2017: 3).

Another issue related to broadcasting and cable service interconnection orders made by TRAI was recently adjudicated by the courts and may be briefly mentioned here. On October 30, 2018, the Supreme Court in *Star India Private Limited vs Department of Industrial Policy and Promotion and Others* upheld the validity of the said orders dating back to 2017. The Supreme Court dismissed an appeal by Star India against the Madras High Court judgment upholding the same. The matter related primarily to the manner in which broadcasters and distributors could package, market and sell television channel subscriptions to end-consumers. Star India had contended that the state governing the TRAI allowed it to only regulate the *means* of transmissions and not the *content* of transmission, as the content was covered and controlled by the Copyright Act, 1956, over which TRAI had no jurisdiction. The Supreme Court, however, disagreed and ruled that both the means and the contents of communications as also the terms and conditions of interconnectivity between the stakeholders could be regulated by TRAI to ensure fair competition and promotion of consumers' interests.

Another matter relates to any conflicts between the issues of regulation and those of competition in the telecom sector. Vikas Kathuria (2018), in a review of theories, maintains that the Competition Commission's intervention is warranted only in "gap" cases where the Regulatory Commission cannot account for consumer welfare. In *Voda Phone India Limited and Others vs Competition Commission of India* (2017), the Bombay High Court observed that

> "unless [the] contract conditions are defined clearly, the rights and obligations of the parties by the Authority under the TRAI Act, the [Competition Commission of India] would not be in a position to decide finally the stated tacit or indirect agreements, by the service providers and its association to settle the progress and/or launching of RJIL [Reliance Jeo India Limited].
>
> (para 101, see also paras 130 (e) to 130 (g))

The court went on to add that the TRAI had the exclusive authority to ensure "quality of service" in the telecom sector.

The Competition Commission of India appealed against this judgment to the Supreme Court. In *Competition Commission of India vs Bhartii Airtel Limited and Others* (2019), the Supreme Court ruled that

> "since the matter pertains to the telecom sector which is specially regulated by the TRAIAct, balance is maintained by permitting TRAI in the first instance to deal with and decide the jurisdictional aspects which can be more competently handled by it. Once that exercise is done and there are findings returned by the TRAI which lead to the *prima facie* conclusion that the IDOs have indulged in anti-competitive practice, the CCI can be activated to investigate the matter going by the criteria laid down in the relevant provisions of the Competition Act and take it to its logical conclusion. This balanced approach in construing the two Acts would take care of Section 60 of the Competition Act as well.
>
> (para 90)

Global and national governance of cyberspace

There is the larger issue of global governance of internet ecosystem and cyber security, which is largely beyond the scope of this paper. The Global Conferences on Cyberspace are held biannually since 2011, where governments, the private sector and civil society organizations assemble to discuss and promote practical cooperation in cyberspace aimed at capacity building and norms of responsible behavior in cyberspace. The first conference was held in November 2011 in London and developed a set of principles for governing behavior in cyberspace. The second conference was held in October 2012 in Budapest to discuss the relationship between internet rights and internet security; the third conference in October 2013 in Seoul developed a framework for and commitment to open and secure Cyberspace. It underlines the principle of universal internet access and parity between the rights that people have offline and ought to have online and reinforces the UN report favoring international law applicable online for maintaining peace and stability and open, secure and accessible ICT environment. The fourth conference was held in April 2015 in the Hague where the establishment of the Global Forum on Cyber Exercise (GFCE) was one of the main outcomes. The fifth conference took place in New Delhi in November 2017, with the focus on the theme of "Secure and Inclusive Cyberspace for Sustainable Development." However, neither an international regulatory regime nor some firm international conventions and laws has yet been firmed up.

In conclusion, it is pertinent to reflect on the way digital technology is impacting the working of traditionally controlled national and international affairs. The national state systems and the international system were based on and have operated through a concentration of powers in the state. Internationalism, supranational regionalism and globalism propelled by the expanding frontiers of digital technology and free trade are significantly changing the state system without making it entirely powerless or extinct. How can governments, multinational corporations and multilateral regional and global agencies act to promote positive behavior in a world of disruptive innovations enabled by ITC?

The Internet rights pioneer John Perry Barlow made a Declaration of Independence of Cyberspace exuding optimism about the power of decentralized networks to push back the more oppressive systems of economic and political governance. For Barlow, the digital realm "is an act of nature [growing] itself through our collective actions." It is a place, he says, where "legal concepts of property, expression, identity, movement, and context do not apply," where "anyone . . . may express his or her beliefs, no matter how singular, without fear of being coerced into silence or conformity," where "governance will emerge" . . . from ethics, enlightened self-interest, and commonweal" (Quoted in Owen, 2015: 206–207).

Digital technologies empower anyone who can build and understand and use them – state as well as non-state actors, ethically as well as unethically. The Canadian media philosopher Owen Taylor observes that the Treaty of Westphalia (1648) ended a century-old phase of international relations among warring empires and created order and stability out of a chaotic and anarchical situation. He goes on to say:

> We face a similar moment today. States as primary units of the international system are being challenged for both power and legitimacy by a wide range of new individuals, groups, and ad hoc networks, all

> empowered by digital technology. While the state has the power to fight back, it does so at the risk of jeopardising the emerging system.

He muses that it is yet to be seen what a new restructuring digitally enabled world would look like (Owen, 2015: 209–210).

The issue of national governance of the cyberspace must, therefore, be constructively reconciled with that of global governance. A great transformation has taken place in the transition from the previous generations of the offline news media and online news media in the latter or last quarter of the 20th century. In the former system, there was a single set of gatekeepers in the form of human editors and reporters. In the new system, the species of news (not views) reporters and editors have tended to become endangered or extinct and have been largely replaced by two sets of creatures, that is, viewsmakers and propagators and editors or mechanical editors without discretionary choices. The search engines like Google and social media platforms like Facebook increasingly rely on automated systems that rank what people see on social media and make decisions on what to display in their newsfeed. To quote Rasmus Kleis Neilsen (2019):

> As people increasingly rely on search engines and social media for news, the information that reaches them has passed through two sets of gatekeepers: First, editors, who decide what to publish, then the algorithms that increasingly shape what we see online.

Reporting on a survey of English language internet users, Neilsen goes on to report that Indians name search engines and social media as their main source of online news, which is strikingly different from other (apparently Western) countries "where people often trust news media more than they trust news found via search engines or social media" Neilsen (2019) concludes:

> Owning, operating, and profiting from algorithms that more and more Indians rely on to find news, gives Google and Facebook great power and great responsibilities – like the human editors who came before them. The 2019 elections will be a decisive test whether they are ready for it.

Notes

1 A previous version of this chapter was published in *Indian Journal of Public Administration*, vol. 64, no. 4, December 2018, pp. 599–613, Special Issue on Independent Regulatory Authorities in India: A Comparative Perspective, Part II. A version of the paper was also delivered at the International Political Science Association (rc-28) conference on "Rethinking Politics, Policy and Governance in Federal Systems: India and the World", University of Delhi, 14–16 November 2013.

2 CERC earlier enjoyed greater autonomy until an amendment to the Electricity Act, 2003, whereby the chairperson as well as its members now "are required to be experts in engineering, law economic, commerce, finance or management" (Section 77 of the 2003 Act). The judicial background of the chair is now not necessary.

3 This plea for representative democratic demand on capitalism must, however, be made in pragmatic rather than ideological spirit. Recently, the Government of India Department of Telecom (DOT) rejected TRAI's recommendation that all service providers be provided with subsidies from the Universal Service Obligation Fund (USOF) for their rural rollouts for a more inclusive telecom policy implementation. USOF, whose present balance has crossed Rs. 15,000 crores, is created out of the payment of 5 percent of their annual revenues by all telcos to this fund. Presently, the DOT subsidizes the highest bidders for setting up additional towers (8,000 in 2007, a proposed 11,000 in 2008) to take mobile services to every part of the country. The TRAI is motivated by providing a level playing field to all companies, whereas the DOT claims that fair competitiveness "has been maintained by giving equal opportunities to all service providers to participate in a transparent bidding process." *The Economic Times*, New Delhi, 22 February 2008: 6.

References

Baijal, Pradip. 2007. 'Let Competition Prevail', *The Times of India*, New Delhi, 30 November, p. 8; A Journey Towards Excellence in Telecommunications: TRAI, 2007. www.trai.gov.in/

Baijal, Pradip. 2008. *Disinvestment in India*, New Delhi: Pearson/Longman, pp. 17–18.

CPR South Young Scholars. n.d. 'Stakeholders' Feedback about the Indian Telecom Regulatory Framework: A Critical Analysis'. www.cprsouth.org/ . . . (accessed on 6 May 2018).

Desai, Ashok V. 2006a. *India's Telecommunications Industry: History, Analysis, Diagnosis*, New Delhi: Sage, pp. 155–215.

Desai, Ashok V. 2006b. *India's Telecommunications Industry: History, Analysis, Diagnosis*, New Delhi: Sage, p. 141.

Desai, Ashok V. 2006c. *India's Telecommunications Industry: History, Analysis, Diagnosis*, New Delhi: Sage, p. 158.

Desai, Ashok V. 2006d. *India's Telecommunications Industry: History, Analysis, Diagnosis*, New Delhi: Sage, pp. 135–136.

Dicey, Albert V. 2008. *An Introduction to the Study of the Law of the Constitution*, Delhi: Universal Law Publishing Co., 10th edition, Indian reprint. First published 1885.

Ghosh, D.N. October 13–19, 2007. 'Regulators Captivated by the Market', *Economic and Political Weekly*, XLII(41): 4094–4095.

Hamilton, Alexander, James Madison and John Jay. 1987. *The Federalist*. Edited with an Introduction and Notes by Max Beloff, New York: Basil Blackwell Ltd., 2nd edition.

India (Republic). 2003a. *Constituent Assembly Debates*, Book No. 4, New Delhi: Lok Sabha Secretariat. 4th reprint, pp. 1644–1666.

India (Republic). 2003b. *Constituent Assembly Debates*, Book Number 2, New Delhi: Lok Sabha Secretariat. 4th reprint, pp. 31–44. Ambedkar's speech as chairman of the drafting committee presenting the draft constitution for debate.

India (Republic). 2003c. *Constituent Assembly Debates*, Book Number 2, New Delhi: Lok Sabha Secretariat, pp. 984–999.

India (Republic). 2009. *Second Administrative Reforms Commission, Thirteenth Report, Organizational Structure of Government of India*, New Delhi: Government of India, Ch. 6.

The Indian Express, New Delhi, 6 December 2017, p. 23.

Kathuria, Vikas. 2018. 'Conflict Between Regulation and Competition Law in the Indian Telecom Sector', *Economic & Political Weekly*, 53(38). www.livemint.com (accessed on 6 February 2018).

Kaur, Kawaljeet and Neena. May–June, 2016. 'Changing Structure, Performance, and Growth of Telecommunications in India', *Arthashastra: Indian Journal of Economics & Research*, 5(3) (accessed on 6 May 2018).

Krishna, R.J. and K. Machad. 2013. 'Telecom sector needs more than FDI', *Wall Street Journal India*. https://blogs.wsj.comindiarealtime/2013 (accessed on 23 September 2020).

Mahaprashasta, Ajay Ashirwad. December 04–17, 2010. 'Fixing Responsibility: The GAG Report . . .', *Frontline*, 27(25): 19.

Mohanty, Bedavyasa and Madhulika Srikumar. 2017. 'TRAI Firmly Anchors Net Neutrality in India, But the Debate is Far from Over: Commentaries', *ORF: Observer Research Foundation*, 1 December, pp. 1–4. www.orfonline.org . . . (accessed on 23 March 2019).

Neilsen, Rasmus Kleis. 2019. 'Algorithm as Editor: Google and Facebook Have Great Responsibility', *The Indian Express*, New Delhi, p. 10.

Owen, Taylor. 2015. *Disruptive Power: The Crisis of the State in the Digital Age*, New York: Oxford University Press, pp. 206–207.

Prasad, R.U.S. 2011a. *Resolving Disputes in Telecommunications: Global Practices and Challenges*, New Delhi: Oxford University Press, Ch. 3.

Prasad, R.U.S. 2011b. *Resolving Disputes in Telecommunications: Global Practices and Challenges*, New Delhi: Oxford University Press, Ch. 3, pp. 74–75.

Prasad, R.U.S. 2011c. *Resolving Disputes in Telecommunications: Global Practices and Challenges*, New Delhi: Oxford University Press, Ch. 3, p. 76.

Sengupta, Arghya. April 9, 2018. 'Facebook's Brave New World', *The Times of India*, New Delhi, p. 12.

Singh, Harsha Vardhan, Anita Soni, and Rajat Kathuria. n.d. 'Telecom Policy Reform in India'. siteresources.worldbank.org/INTRANETTRADE/Resources/Singh.pdf (accessed on 6 May 2018).

Sukumar, Arun Mohan. 2015. 'Another Blow to the Autonomy of Telecom Regulator', *The Wire*, 30 July. https://thewire.in/politics/nda-appoints -one of-its-own-as-new-trai-chairperson (accessed on 6 May 2018).

Supreme Court. 2018. *Competition Commission of India v. Bharti Airtel Limited and Others*, New Delhi, 5 December. Reportable PDF. www.sci.gov.in (accessed on 25 March 2019).

Supreme Court. *Star India Private Limited vs Department of Industrial Policy and Promotion and Others*. https://indiankanoon.org (accessed on 24 March 2019).

Telecom Regulatory Authority of India, *Annual Report 2015–16: 86–87*.

The Telecom Regulatory Authority of India Act, 1997, Section 11.

The Times of India, New Delhi, 9 February 2018, p. 1.

4

DELHI AND STATEHOOD

Problematic law and order and federal trade-off

Amna Mirza and Bhav Nath Jha

Since the early 1950s, the issue of statehood for Delhi has witnessed interesting arguments and twists. Power is essential for any political governance mechanism wherein the overlapping of the domain of authority, or lack of it, can hamper the working of an elected government. This has been echoed in government chambers of Delhi on several occasions, and often the need to make Delhi a state was put forward as a panacea for ailments with regard to governing the capital. Delhi, at present, is a union territory with special status. Unlike other states, in the case of Delhi, issues concerning law, order and land fall in the ambit of the central government. Further, as Article 239AA of the Constitution states, the lieutenant governor (LG), appointed by the president, is the administrator of the National Capital Territory of Delhi. This chapter explores the trajectory of debates on issues surrounding statehood for Delhi through insights into history, judicial pronouncements and reflections from a contemporary political scenario, among others.

Delhi, officially the National Capital Territory of Delhi, is a federally administered union territory of India. It is a cosmopolitan city. When we look at the geographical contours, Delhi has an area of 1,483 sq. km surrounded by two neighboring states: Haryana and Uttar Pradesh. Moving away from situational aspects, the city has a rich history. Delhi has witnessed the rise and fall of several empires, such as the Mughal, British, etc. It shall be seen in later parts of this chapter how historical and geographical inputs have made contemporary governance in Delhi a contested phenomenon.

Time and again, the demand for "statehood" has been raised. It has been the electoral promise for several political parties in the run-up to Delhi Assembly elections many times. The 2015 Assembly relations witnessed heated arguments over this issue again, which saw a "State of Delhi Bill" emerging in 2016. In fact, the demand for full statehood for Delhi is not new. This issue was also raised in Constituent Assembly Debates. The ruling Aam Aadmi Party (AAP) has been continuously demanding full statehood for Delhi. It has been one of the essential elements of the election manifesto of the Aam Aadmi Party, as well.[1] But there is no consensus among the stakeholders on this issue. The lack of political will and other administrative constraints have also made the demand for full statehood for Delhi far from reality.

After a long struggle, India attained freedom in from colonial rule in 1947 and adopted a federal model of governance in which division of powers is one of the basic principles of the constitution, as power is divided between the union government and state governments under Article 246 and the Seventh Schedule of the Indian constitution. But the Indian federal system is asymmetrical. There are some special provisions in the constitution contrary to the classical model of the United States of America. For example, Delhi is a union territory. Land, police and public order are under the direct control of the central government, whereas health, education, water, electricity, etc., come under the jurisdiction of the government of the NCT of Delhi. Thus, the situation is very complex, which leads to clash and confusion between the two levels of government. It is a healthy sign of democracy that such issues are being raised. However, one cannot neglect the politically astute game plan going alongside.

History as locus of power

Delhi has been in focus for power and domination in several phases of history. From being capital of the reign of Iltutmish, to being the fulcrum of several activities during the first war of independence in the 1857 revolt, Delhi has on several occasions waded through troubled waters. The capital of colonial India earlier was Calcutta (now Kolkata), which was later shifted to Delhi in the year 1911. If we go by the provisions of the Government of India Act 1919, and later 1935, Delhi was designated as a central government administered province (Mukherjee: 2009). As the cries and struggle for independence gained momentum, these attempts were made to understand shortage and loopholes in administration of units, and here it is pertinent to note that the Sitaramayya committee in the year 1947 also remarked on constitutional and administrative processes that were at loggerheads with development in Delhi owing to the absence of proper coordination among various agencies and units. However, it refrained from the idea of granting full statehood to Delhi. The Constituent Assembly also took a similar stand on this issue. If we go by the records of history, the erstwhile Jana Sangh was not in favor of demand for statehood at that time (Kumar: 1991).

In 1952, Chaudhary Brahm Prakash became the first chief minister of Delhi, with a directly elected legislative assembly. However, it did not last long. With the emergence of the State Reorganisation Commission (SRC) in 1953, there were states and union territories (UTs), and Delhi was placed as a UT. The municipal corporation of Delhi came into existence in 1958 to cater to essential services for people of Delhi and also work in tandem with several central government agencies. The legislative assembly was abolished by Constitution (Seventh Amendment) Act, 1956, in order to implement the recommendations of SRC. Delhi was formally declared the National Capital Territory after the Constitution (Sixty-ninth Amendment) Act, 1991, and subsequently the Legislative Assembly was reestablished in 1993.

The Jana Sangh Committee remarked that granting of full statehood of Delhi would not be good for the interest of the country as a whole. Later, with time, the Prabhu Committee came up in 1975, which made certain recommendations in order to facilitate greater decentralization of power and to situate the administration of Delhi closer to the people, avoiding a multiplicity of power centers and the delegation of powers, among others, were key recommendations of this committee, but there was hardly any action on these. In 1987, it was pointed again by the Delhi Congress Party that the administration in Delhi was far away from the needs of the people of Delhi. The administration was not accountable as well as not responsible for the democratic mandate of the elected bodies in Delhi. The governance in such scenarios was reduced to red-tapism, delay in work and bureaucratic handles, among others. At that time, with Prime Minister Rajiv Gandhi, the Committee for Reorganisation of Delhi was set up, and after all deliberation, made a point for Delhi to become a union territory, which also made way for the union government's interference. After the passing of the 74th Constitution Amendment Bill in 1992, Article 239AA and 239AB assigned the territory of Delhi to the LG and mandated the LG to act on the aid and advice of the chief minister and his council of ministers. The powers and functions of the LG were described, and it has been pointed out and argued that clause (4) of Article 239AA has been an axis of dispute in a parliamentary democracy like India. The LG, who does not owe his authority directly to the popular mandate, was the constitutional head of the UT and also the prime decision maker. This clause has also witnessed misuse by the change of regimes, when two opposing parties are in power at the central and the UT level in Delhi, which has led to skirmishes between the two.

As for the recent record, the AAP government in Delhi has upheld its promise, and the issue of full statehood for Delhi has been put up for discussion in the Delhi Assembly session. There are experts who find merit in this case and support the need for more powers to the state government. This also is in tune with the practical needs of contemporary times, which demand amendment and changes in the current status of Delhi.

Upon further discussion of the matter, certain practical insights demand attention: namely that Delhi's government does not have to remain in the terror of transfer and posting of its officials, that control of the Lutyens' zone is with the union government and that the New Delhi Municipal Corporation is delineated from the rest of the area of Delhi. However, this initiative has been scuttled by the current opposition party in Delhi, the BJP, possibly as an attempt to divert attention from its nonperformance in the union government.

No debate in governance is devoid of political dimensions. In the same vein, it has been alleged by the current Delhi government that the union government has turned a blind eye to its demand of full statehood ever since current Prime Minister Narendra Modi came to power. This was different in the past as former Chief Minister Madan Lal Khurana and Sahib Singh Verma had supported full statehood for Delhi. It was in 2003 that then Deputy Prime Minister L.K. Advani presented the Delhi Statehood Bill.

It goes without saying that the crux of the jurisdictional crisis in Delhi is attributed to Article 239AA of the Indian Constitution. The LG can disagree with many decisions of elected representatives. The executive power of the council of ministers is limited in this regard. This puts the elected government in a situation of *responsibility without power*.

In a recent judgment, the Delhi High Court has pronounced the verdict which maintains that Delhi is a union territory. It ruled that the lieutenant governor was the administrative head of Delhi, giving him autonomy and necessitating that all decisions go through his office (Singh :2016). This verdict leaves several issues to ponder over for debate in the governance model of Delhi; police with reference to law and order has a special bearing in this regard, as it leaves many unanswered questions. The constitutional status of the National Capital Region, under Article 239AA of the Constitution, does not cut in the effect of Article 239, which entitles the LG to act independently of his council of ministers. This brings into focus the federal debate over the mandatory concurrence of a representative of the center/union, that is, the lieutenant governor and peoples' elected representatives of Delhi being essential for administrative decisions.

The contemporary political tussle, a baggage of the past, problematizing the constitutional texts with reference to issues of the present has indeed opened a topic for deliberation whether the right balance has been struck between union government responsibility and the democratic principle that people are ruled by a representative government.

The structures and political apparatus today are open to multiple interpretations about who really runs what aspect of the administration. One of the most contentious issues is law and order. Delhi is a union territory; hence, the law and order here is under the central government, while the ruling AAP believes that the people of Delhi have elected them for their safety and prosperity. In fact, it is a matter of consideration that you can't hold the elected government of the state accountable for law and order. Therefore, it is said that such problems can be solved only if Delhi gets full statehood status. However, ambiguous legal and constitutional reality comes in to halt this solution.

Further, as we reflect on Supreme Court judgment that debunks the existing power structure and says that LG is not the permanent executive, because, being an agent or nominee of the center, the position can be changed by discretion of the center at any moment (**The Hindu:2016**). What comes forward is a complex web – where the power tussle continues and governance, due to the federal debate, becomes difficult. It is argued that Delhi should not be under the control of the Delhi government because it could mean unnecessary administrative and political interference in the name of efficient governance and also because disturbing the existing set-up of Delhi police could lead to complete chaos.

It is not that the idea of two police forces operating within a large metropolis lacks precedent, as it exists in London, too. However, fears of politicization of the administration loom large. Political leadership is a transient phenomenon. This is why neutrality of administration, law and order, is the best way to retain its judiciousness and a means to guard sanity. Bureaucracy is an antidote to partisan politics, and this mutual confidence paved the way for a healthy government dialogue and work mechanism.

One should keep in mind that Delhi is not only a city-state but also the capital of India (Slack and Chattopadhayay: 2009 & 2011). All the important establishments of the central governments are in Delhi. So the central government requires significant departments such as police, land and public order for the management of its affairs (Saxena: 2017). Besides this argument, a constitutional amendment is desirable for putting Delhi police under the state government. Time and again, there is a notion that the party winning the Delhi Legislative Assembly elections does not have the administrative prowess to run the territory, particularly because the police do not report to it.

Law and order is controlled by the union government. No doubt, governance becomes a victim in this strange structure, and even the working of the police has been hampered due to duplication of efforts several times.[2] This raging turf war with the union government even led to demand for referendum on lines of Brexit (Hindustan Times: 2016). No one can dispute that Delhi, with the legislative assembly, is a unique example of asymmetrical federalism in India. However, stirring up issues of police for statehood is pertinent to the popular political bargaining for arousing citizen imagination. It cannot be seen as an immature furor but has to be seriously considered for the sake of unity of the nation and sanctity of the Constitution.[3]

The political matrix continues with the blame game.

> Congress has worked effectively, balancing all issues, irrespective of our regime being in power in the Centre or not. What can be seen in current times is that the union government is trying to hoodwink the people of Delhi on the issue of full statehood.[4]

Another important caveat in the debate is that the change in the constitutional status of Delhi must not create anarchy. Respect for procedural democracy in sense of legal discourse is essential.

Fractured jurisdiction

Delhi presents a unique scenario of governance – where the central government manages a union territory and at the same time there is the existence of a state government which is elected by the people. The legislative power of land and public order are with the union government. All of Delhi's laws are under control of the Delhi Development Authority (DDA), and public order is not in the state government's ambit, as the union Home Ministry is responsible for the Delhi Police. A country's capital is the landmark of its government and administration. Political memberships have led to turbulence. Experts also opine that the political class may change colors for its vested agenda, but they fail to grasp the seriousness of the reality. There will be an adverse impact, and a reign of anarchy shall prevail if Delhi is granted statehood. On the ideological plank, it is argued that Delhi belongs to the people of the country. Its outreach, which is pan-Indian, cannot be reduced or issues cannot be resolved by mere reduction of complexity of problems with achievement of statehood. Delhi's ethos cannot be patented by any one dogma as its sole custodian or some of the larger ambit of life in vicinity of Delhi cannot be seen in parochial sentiment in terms of attachment to land. There are several diplomatic issues at stake, and other security matters will also bear the brunt of this bifurcation. As the tussle continues, the need of the hour is a proper coordination between the governments for the all-around development of this union territory and the country as well.

With this fractured jurisdiction and further deflection of blame (between the union and the state governments) when it comes to law and order, it is high time that for our burgeoning capital, we realize and ensure reforms, because it has not been policed effectively by the Ministry of Home Affairs, to which an increase in crime in the capital region bears testimony. Politics and issues of polity thereof are often contested. The churnings of politics also has impacted the issue of statehood for Delhi. Changes in context often lead to transformation in narrative, and a similar thing was seen in how AAP did not put the emphasis on full statehood for Delhi, as it should have been. Amidst the debate of one nation, one poll, there were critics who pointed out that one must not dismiss the maturity of the voter who can distinguish between issues of national and state polls. When the drill for Delhi elections was going on in January 2020, held against the backdrop of a giant mandate to BJP at the center, with mere token words for statehood, the commitment to the erstwhile pledge for full statehood was seen to be losing its strength by the then-incumbent Chief Minister Arvind Kejriwal.

What was stranger for political observers is that though AAP is a political novice, yet even on the part of BJP, there was a conscious distance in terms of support of the issue of statehood for Delhi, which the party had dearly fought for since the 1980s. Some also remarked that the elections in Delhi in 2020 were contested on issues of governance (Kulkarni : 2020). Improvements in drinking water, educational infrastructure and cheap yet better quality in health care, among others, were main issues highlighted in the manifesto of AAP, and using this plank, the vote was sought to continue with development works. Further, the Delhi model was presented to the voters, which rallied around the idea of the chief minister as an able administrator.

Election exercise witnessed in 2020 in Delhi saw how statehood found scarce relevance in manifesto, political speeches, which has been interpreted by analyst as an attempt to undo anarchist perception associated with the incumbent party in Delhi, with certain sections of voters. With changes in the demographic profile, Delhi, with mostly inhabitants from many states being voters here, did lead to a dilution of debates on statehood during the election campaign, too. Statehood was not seen viable as an issue as a vote gathering enabler and election winning input owing to changes in context of political narrative at the National level. One must also do a comparative analysis here that, unlike other states such as Uttarakhand, Chhattisgarh or pleas for carving out Vidharbha or Bhilistan, the efforts and voice of solidarity has been largely from the people and grassroots elements. This democratic movement has been conspicuous by its absence in case of statehood for Delhi.

The political saga on any issue delves within emotive and feasibility aspects. The issue of Delhi statehood may have been zealously voiced by various parties, yet several commentaries in Delhi polls also showed that there was lack of enthusiasm for it among the people residing in Delhi. This lack of enthusiasm for statehood among them can be attributed to people's comprehension of the issues as nonviable owing to the long procedure for amending the law of land without guarantee that good results will flow once statehood is achieved, or apathy due to the need for quick results of government machinery amid the pacing world outside. It can be even said that what made the debate on statehood become less relevant for the political landscape of Delhi is that people perceived it as an excuse for elected representatives to shirk duty.

Nonetheless, the issue of statehood has been seen to be revived to pin down the shifting arena of blame games in case of any lapse, even after the polls. With the lack of interest for the issue, its electoral prospects are diminished, or lack of feasibility might cast doubts for the future of the issue, yet one cannot dismiss the need to look at a twin-government structure in the same vicinity, which impacts the lives of people. Democratic deliberations with all stakeholders must always continue to let the idea of power-sharing ushered by federalism yield benefits of governance in Delhi.

Notes

1 We are grateful to Professor Rekha Saxena, whose insightful and in-depth study of this issue has played a vital role in shaping the arguments of this chapter.
2 Interaction with Mr. Abdul Siddiqui, former DIG Police Punjab, New Delhi, 2 September 2016.
3 Interaction with a senior functionary of BJP, New Delhi, 7 September 2016.

4 Interaction with Delhi Congress leaders, and representatives of Aam Aadmi Party, New Delhi, between 19 August 2016 and 3 September 2016.

References

The Constitution of India. 2006. *Allahabad: Third Edition*, Central Law Publications.

Johnson, D. and Richard Wilson. 2015. *The Last Imperial City*, London: Palgrave McMillan.

Hindustan Times , 'After Brexit, CM Kejriwal Calls for Referendum on Delhi Statehood', , New Delhi, June 24, 2016.

Kulkarni, Sagar.. 'Why Statehood Issue is off the Table in Delhi Elections 2020', *Deccan Herald*, January 28, 2020.

www.google.com/amp/s/www.deccanherald.com/amp/opinion/why-statehood-issue-is-off-the-table-in-delhi-elections-2020-798627.html

Kumar, Sudhir. 1991. *Political and Administrative Setup of Union Territories in India*, New Delhi: Mittal Publications.

Mukherjee, Mithi. 2009. *India in the Shadows of Empire: A Legal and Political History (1774–1950)*, New Delhi: Oxford University Press.

Saxena, Rekha. 2017. 'Multilevel Framework of Governance in Delhi: A Non-Sovereign "State"', in Himanshu Roy, M.P. Singh and A.P.S. Chouhan (eds.), *State Politics in India*, New Delhi: Primus Books.

Singh, Soibam Rocky. 5 August, 2016. 'L-G Najeeb Jung Is the Boss in Delhi: High Court Tells Kejriwal's AAP Govt', *Hindustan Times*, New Delhi.

Slack, Enid and Rupak Chattopadhyay (eds.). *Finance and Governance of Capital Cities in Federal Systems*. London: McGill Queen's University Press, 2009.

Slack, Enid and Rupak Chattopadhyay. "Governance of Capital Cities in Federal Countries: 18 Comparative Perspectives." In *Varieties of Federal Governance: Major Contemporary Models*, edited by Rekha Saxena, p.p:18-34, Delhi: Foundation, Cambridge University Press, India, 2011.

'Supreme Court Refuses to Stay Delhi HC Verdict Declaring Delhi a Union Territory', *The Hindu*, New Delhi, September 9, 2016.

Part II

ENVIRONMENTAL AND RESOURCE FEDERALISM

5

LAND ACQUISITION, MOVEMENT ACTORS AND FEDERALISM

The anti-POSCO movement in Odisha

Biswajit Mohanty and Babita Verma

Federalism, which is very much a part of the democratic set-up because of its stress on the separation of power, is generally understood as layered negotiations for sharing of power among actors at multiple levels. The mainstream theories of federalism advocate how best power and responsibility can be shared among different levels of governments so that financial and administrative goals are achieved efficiently. As nation-states matured, political and economic processes became more complex, especially in the era of globalization, and various institutional and non-state actors are entangled within the web of federated institutions. The competitive spirit unleashed had various effects on the federating units and political actors. Some actors are accommodated, and others were abandoned to fend for themselves, resulting in the growth of political resistance. New frames of analyses tried to understand the changing dynamics. In this context, the fundamental questions to be answered are: how would federalism as an analytical tool be applied to understand the issues brought up by social movements? As praxis, how has federalism fared to provide room for the new political actors within its fold to resolve issues and accommodate demands?

The normative economic theory of federalism vouches for "fiscal equivalence" through the successful distribution of power among different tiers of governance; land and resource federalism, in contrast, stresses on the land resource management through the procedural governance of "right to notice," "right to be heard" and "right to appeal" by the stakeholders within a formal set-up. Multilevel federal governance, which is defined as "the sum of regulations brought about by actors, processes as well as structures," stresses the negotiations and engagements of institutionalized actors in deliberating, formulating and implementing policies (Walti, 2010: p. 2). It does not take into account people's struggle to retain private and common property resources on which their livelihood depends. They have been denied historical rights that they have been enjoying since time immemorial in order to promote the national interest. Except for the multi-level governance framework in a very limited way, none of these federal theories analyze the issue of agency of movement actors in a federal democratic process.

The reasons for not taking the social movements and issues into the analysis are the following. The first reason is the excessive formalism of the theories in analyzing the relation among the constitutional units. The second is the unnecessary stress put on fiscal devolution and constitutional power-sharing processes. The third problem is the deliberate oversight by the theorists of the existing complex power structure of polity that excludes people from the decision-making process to decide about their common good. Finally, it is the failure of the federal theorist in ignoring the domination of certain non-state actors at the cost of other stakeholders in resource sharing. These four reasons have prompted the leading theorists of federalism to overlook popular movements, which are generally categorized as employing a non-constitutional method of demanding their share. In this context, the chapter, taking the anti-POSCO movement in Odisha as the case study, argues how federalism in India stresses territoriality and distribution of power among units and neglects the real issue of people's empowerment and participation as the substantive core of democratic federal practice. The chapter further argues that the movement is not just a success story but also a critique of contemporary federalism as a theory and practice.

The reasons for selecting the case study are: first, it is an interesting study to understand how different sets of actors within a given federal structure have been competing to secure their respective interests. On the one hand, there is the movement, which is trying to protect peoples' livelihood through organized struggle. On the other hand, POSCO, the multinational company, is endeavoring to secure its interest to establish a Special Economic Zone purely for private gains. There is a third set of players – the three tiers of government – trying to negotiate between them for their legitimacy within the federal rules of governance. The second reason for selecting the case study is to observe when and how multilevel governance fares or fails in practice with the presence of social and political movements. The third reason for taking up the case study is to examine how movement actors maneuver, negotiate and innovatively utilize their limited resources to influence institutions and restructure the federal power relations. In other words, the study attempts to unravel the complex dynamic relationship that exists among formal federative structures, internal agents of change and external dominant non-state actors.

Theories of federalism and social movement actors

The traditional theories of federalism have not been able to incorporate the movements within their framework satisfactorily. For instance, fiscal federalism, which is primarily normative, in nature would suggest that allocation of resources and functions between levels of government affects policy performance. Economic federalism also stresses that for policies to be successful, the distribution of resources must conform to the principle of "fiscal equivalence" which would be achieved through Pareto optimality in a market society (Walti, 2010). The presence of competitive party politics, many a time, upsets the market arrangements and restructures the federal arrangement. In this context the question is: should resource distribution be carried out according to the market principles or government-led distributive mechanisms? Fiscal federalism in essence talks of the market mechanism to curb government overreach to achieve public good (Walti, 2013). Logically, it glosses over the institutions of governance and social and political actors' role in transforming federal practices or influencing fiscal policies. One such political actor is the social movements that are part of the competitive politics demanding just and equitable distribution of financial and natural resources. Where do movement actors fit within the theories of federalism?

Multilevel governance (MLG) and the actor-centered theories have innovatively incorporated various actors within their framework. According to Hooghe and Marks, there are two kinds of MLGs – MLG-I and MLG-II. The MLG-II type takes into account the actors whose role becomes important in federal politics (Hooghe and Marks, 2010). Though it is difficult to compartmentalize the two types of MLGs because of the presence of abundant institutions, the MLG-II takes cognizance of the presence of actors in a federal polity that intersects and competes for distribution of power and resources.

There are different kinds of actors in a polity. The actors, generally speaking, are the "composite of individuals, firms or associations" in the policy-making arena; those constitute organized political actors based on sociocultural issues, citizenship rights, political and economic interests and religious and ethnic communities (Hassel, 2010: p. 154). There are four types of collective actors: clubs, associations, social movements and coalitions, according to Hassel (for details see Hassel, 2010: p. 155). Their roles in pursuing the public good and influencing policy formulation, evaluation and implementation vary. But it is debatable whether all kinds of social movements are part of the civil society or a particular kind is part of it.

Hassel distinguishes between different kinds of civil society actors based on their role in the federal polity. The roles of CSAs (civil society actors) and corporate interests are different. Corporate interests believe in narrow economic gains and profits. Their eagerness to contribute resources, management skills and logistics to solve the problems are dependent on the maximization of their interests. CSAs, in contrast, are a necessary part of the democratic discourses and provide "alternatives and legitimate policy positions." The other kind of distinction that has been made between the two is in the sphere of agenda-setting. The companies and other private actors gain easy "institutionalised/centralised" access to the policy-making processes through different associations and club membership, through their involvement in advisory committees, social and economic councils. The social movements and their coalitions have no formal avenues within the political institutions to gain access to authorities to impress upon and influence them except by using "voice strategies," such as demonstrations, public opinion campaigns, media events and so on (Hassel, 2010: p. 156). The major problem is that the MLG restricts the CSAs to the confines of the formalized and institutionalized groups, for instance, party, recognized interest and pressure groups and registered NGOs. The political movement actors or extra- or non-constitutional actors do play an important role in critiquing the government policy and challenge the very power structure of the federal set up.

In the context of the distinction between the different types of multilevel governance (MLG), different types of organized interests can be expected to coincide with different types of MLGs (Hassel, 2010: p. 156). In this schema, civil society actors could be put into the MLG-II matrix for the following reasons. First, doing so supports a very "complex" and "fluid" patchwork of overlapping jurisdictions of federal units that attracts more flexible interest groups into its fold. Since the groups are driven by ad hoc policy issues, weaker organizational structure and underdeveloped internal decision-making processes, they find it easier to function with the governments at different levels. Second, the MLG structure is capable of providing solutions to new societal problems through innovative public policy responses. And, finally, this puts the importance of the "voice strategy" role enacted by the fluid informal groups in bringing policy issues to the authorities. Because they have less political access to formal institutions, they –through mobilizing public opinion, strong demonstrations tactics and blockades – attract media visibility and public gaze. Through the voice strategy, they are mostly successful in forcing the institutions to become "veto points" or opportunity structures for the movement actors. Sometimes they fail, and other times they succeed. The failure may be due to lack of political and economic resources to influence agenda-setting in a political system (Hassel, 2010: pp. 156–157). The agenda-setting role may be an important aspect of a civil society actor, but given the structural relation within a polity, it is difficult to gauge what kind of social movements and actors succeed and in what given structural conditions agenda setting can be reversed.

Land and resource federalism is an important analytical framework to understand federalism in the era of globalization. It generally focuses on institutional practices for sharing resources. In the era of globalization, salience of "local" has increased due to its territorial overreach; so also have conflicting claims over local natural resources. Economic globalization has changed the nature of territorial governance from top-down division of power to the bottom-up approach, where local political and other actors have taken control over their political, social, cultural and economic affairs (Loughlin, 2013).

Land has become an important commodity, especially in the post-liberalization period, for establishing Special Economic Zones, building infrastructure and development projects, and capital investment in realty sectors, theme parks, IT hubs and smart cities/zones. As the claims over land increased, the most contentious issue of acquisition of land for "public purpose" took a political turn. Compulsory land acquisition required striking a balance between land acquired for public purposes and land security for the peasants along with the protection of the right to private property and community-owned common resources. It was argued by the movement activists that while acquiring land, the "procedural rights" – "the right of notice," "the right to be heard" and "the right to appeal" – should be strictly adhered to. The procedure must include transparency of decision making, consultation with stakeholders and fair compensation to the displaced and project-affected families (FAO, 2008, p. 5).

Land and resource federalism addresses, first, the constitutional construct of resources and its interpretation through legislative provisions; second, explores the administrative, financial and institutional linkages between the center and the states; and third, examines asymmetry in resource distribution and relates it to the concerns of federal functioning (Chaturvedi, 2016: p. 4; Noronha et al., 2009). For instance, the constitutional provision on land in India is derived from the Article 246, where the land-use pattern falls within the state list and the states have "rights in or over land," "land tenures," "including transfer and alienation of agricultural land" and "colonisation," whereas the existing provision for acquisition and requisition of private land by the government is provided in the concurrent list. The land, in terms of ownership or property, is dealt with Article 31(1) of the Constitution. But subsequently, the right to property was deleted and 31(1) was inserted in the constitution as 300A under the provision of "Eminent Domain." By making the right to property a constitutional and legal right, the Constitution gave leeway to the legislature to restrict the person's right to property and ensured that the person is entitled to fair compensation. The demand of the movement not to part with their resources and property could not fit into the distributive aspect of federalism. Resource federalism hardly took cognizance of the role of what it considered extra-constitutional actors such as social movements and of the rights of people who contest the claim of the state over their private property and common property resources.

This chapter, taking the case study of the anti-POSCO movement, argues how people's movements have challenged the principles of federalism propounded by traditional theories. Though the MLG model has limitations, we have adopted it as the frame of analysis, with slight modification, as it provides the space for political actors in its analytical framework.

The anti-POSCO movement

On June 22, 2005, a memorandum of understanding (MoU) was signed between the POHANG steel authority of Korea and the Odisha government to set up an integrated steel plant in the Jagatsinghpur district that would produce 12 million tons of steel per annum. The MoU signed had three components spread across three districts of Odisha: first, a steel plant and a captive port at Paradeep would be set up in the Jagatsinghpur district; second, mineral supply to the plant would be undertaken from the Khandadhar mining area spread over the Keonjhar and Sundergarh district; and third, extraction of water would be done from the Mahanadi from the Cuttack district. In the MoU it was further mentioned that the company would require 4,000 acres of land for the purpose to set up the steel project and associated facilities, which included the port facilities.

In addition to this demand for land, the company was to acquire 2,000 acres of land for developing the township and 1,500 acres of land near the port, and the rest would be near the mining area. The MoU also mentioned that the government would provide the company the transportation facilities, an uninterrupted water supply for the project and other project-related ancillary infrastructural requirements. The Industrial Development Corporation (IDCO) of the government of Odisha would acquire land. The rationale for establishing the company in Odisha was: (a) willingness of the government of Odisha for having foreign direct investment (FDI); (b) the availability of cheap mineral ore, suitable site for port, water facility and surplus electricity; and (c) the economic benefit that would accrue to the state in terms of its contribution to Odisha's economic gains.[1] This claim has been systematically demolished by the Mining Solidarity Zone report 2010. (Mining Solidarity Zone, 2010)

Gadkujang, Nuagaon and Dhinkia were the three *panchayats* adversely affected by the project. The total land to be acquired was 9,829 acres coming from eight villages – Polang (766 acres), Bhuinyapal (981), Bayanalkandha (321 acres), Noliashahi (70 acres), Nuagaon (1,932 acres), Jatadhar (2,352 acres), Dhinkia (2,110 acres) and Govindpur (1,297 acres). Out of the total land required, 89 percent would be government land and 11 percent would be private land. According to the village leaders, on average, in the three *panchayats*, the landed property holders constituted about 10 percent, and the rest are landless. The landed class also were owners of large betel vines, a lucrative cash crop, where the landless people work for a living in the off seasons. According to a report, there are about 5,000 betel vines that employed around 10,000 people round the year (National Center for Advocacy Studies or NCAS, 2010; Mishra and Nayak, 2011). The recent field visit in 2016 revealed that the number of vines has increased, as people are borrowing from the banks to cultivate betel on the forest land. These vines are employing women and elderly persons. Earnings from this vine range from a hundred thousand rupees to a thousand rupees, depending on the size of the vine. People also grow cashews, rice and drumsticks to supplement their earnings.

The other worst project-affected people are Nolias, the traditional fisherfolk. Their income from fishing in the river and sea ranges from 100 to 5,000 rupees a day. The fisherfolk were never counted as the directly affected nor project-affected people by the POSCO and the Odisha government (NCAS, 2010). Seeing the imminent danger of loss of livelihood, the fisherfolk were at the forefront of the movement against the POSCO.

The opposition to the project came from various environmental activists (Roy and Dey, 2013) who opposed the construction of the port at the Jatadhari estuary on the grounds that once the Jatadhari port was constructed, there would be restriction of the flow of freshwater into the sea that would cause water logging in the nearby areas. Drawing water from the Jobra barrage at the rate of 75,000 cubic meters every day (according to a letter from the company to the state government dated 28 July 2005) for industrial purposes would have an adverse effect on agriculture and the daily water needs of Cuttack, Jagatsinghpur, Jajpur, Khurda and Kendrapara. People in Cuttack formed Mahanadi Surakshya Samiti to oppose POSCO, which has plans for lifting water from the Zobra Barrage for commercial purpose.

The POSCO Pratirodh Sangram Samiti (PPSS), the vanguard organization at the local level, opposed the project by mobilizing and organizing the local people, political parties, civil society actors and advocacy groups to stall the project. Abhay Sahoo, a veteran leader of the Communist Party of India, provided leadership, organized villagers and carried out a sustained and powerful anti-POSCO movement. There have been several human rights violations in the form of killing people, torturing and beating women and children, and incarcerating the activists and leaders (International Human Rights Clinic, ESCR-Net, 2013; PUCL Report, 2011). After a prolonged struggle, the movement was successful when the company declared in July 2015 its final withdrawal from India.

What went wrong?

The first phase (2005–2010) of the project saw perfect self-rule and shared rule where the subunits had the autonomy to decide about the project. There was also a perfect coupling between the central government and the state unit. The Congress party at the center and the Biju Janata Dal in Odisha were on the same side of the fence as far as the project was concerned. They also agreed on the flow of FDI to the Odisha government. POSCO was also closely monitored by the PMO's, office as it constituted the highest ever FDI. This was the period of cooperative federalism practice to implement the project. The evidences of the existence of cooperative federalism are:

a According to the MoU provisions, the state government was obliged to procure mandatory environmental clearance from the central government as the project was coming up in the Coastal Regulation Zone (CRZ). As per the records, the clearance to the port was given by the Environment Ministry of India on 15 May 2007 along with the Environment Clearance for the captive port despite severe protest by the PPSS. Final clearance to cut 2.8 lakh trees in the forest land was given "in principle" on 29 December 2009 (Meena Gupta Committee Report, 2008).

b Of the total land that the company had procured, 74 percent constituted forest land for non-forest use. Under this circumstance, where 2.8 lakh trees would be cut, the Forest (Conservation) Act 1980 would automatically be invoked, and clearance should be sought from the concerned department. Despite the Supreme Court order, the clearance from the central government was obtained "in principle" based on the advice of the Forest Advisory Committee. The project was finally allowed to operate on Dec 29, 2009 (Meena Gupta Committee Report, 2008). The Environment Ministry allowed the state government to hand over 2959 acres of forest land to POSCO (The Economic Times, 2010).

c *The Hindu* reported that Naveen Pattanaik, the Chief Minister of Odisha, had come to meet President Lee, the President of South Korea, at the dinner hosted by the Indian president. He told the media persons that "the State government would speed up land acquisition for the project," which has not taken off due to stiff resistance to the project by locals (*The Hindu*, 26 January 2010).

d The center went as far as to order the dismissal of the Sarpanch of Dhinkia *panchayat* to effortlessly carry out the land acquisition and advised the Navin Patnaik government to change provisions of the MoU in the light of recent developments. Jairam Ramesh, the then Minister of MoEF, on 2 May 2011 wrote back to the Odisha government, in response to the state government's letter on holding of the *gram sabha* following the Pachayat Raj Act of 1964 and Forest Rights Act of 2006. He wrote that he reposes full faith in the Odisha government's report and that he "respects the reports from the SDO and the Collector." Therefore, he clarified that there is no justification of claims made by the Dhinkia and Gobindpur residents on recognition of rights under the Forest Rights Act, 2006. The reposition of faith, he stressed, is the "pillar of cooperative federalism." He also stated in the letter that the trustworthiness of the elected government should not always be questioned by the central government (Letter by Jairam Ramesh, 2011).

The local government also helped the state and the central government to achieve the desired goal. Despite all the support from the center, why did the project not take off? What went wrong that forced the biggest FDI to withdraw? What does it mean for federal practices?

Two plausible answers seem to be relevant. First, the rivalry between the BJD and the Congress is one of the probable reasons of stalling of the project. The second reason is the presence of a strong movement despite severe repression, killings, registering false charges and several human rights violations that delayed land acquisition, implementation and final withdrawal of the project (Interview Abhay Sahoo, at Ersama on 9 March 2016). According to Abhay Sahoo, the movement not only challenged the existing federal structure but questioned the very basis of nondevolution of power under different garb, where the lowest unit of the federal structure was disempowered. This can be ascertained from the following facts.

i Lack of focus on the principle of subsidiarity by federal units

With the enactment of the 73rd Amendment Act, multilevel governance witnessed a new structural arrangement of power, where local governments became autonomous and were conceived to be acting as equal partners of development. The federal relationship, instead of falling into the hierarchical center and state relationship, had to operate along with the local. The 73rd Amendment Act, by giving power to *gram panchayat* to decide about their livelihood issues, which includes deciding to efficiently manage their resources (land, in this case), paved the way for sustainable development. The 11th schedule of the constitution provides for powers to the *panchayats*: to improve land quality, implement land reforms, carry out land consolidation and conserve soil. The 11th schedule may not be very effective and significant in itself as far as land acquisition goes, but its significance can be gauged when one reads it along with the Forest Rights Act, 2006, and environmental laws.

The Scheduled Tribes and Other Traditional Forest-Dwellers Recognition of Forest Rights Act, commonly known as the FRA 2006, was enacted to reverse "historical injustice" handed out to the vulnerable tribes and OTFDs (Other Traditional Forest Dwellers) who have been considered as an integral part of the forest ecosystem on which they survive and sustain the system in the liberalization and globalization period in India (FRA, 2006). Though the FRA is clear about the goal, there was much illegality of giving clearance to POSCO in violation of rules of the act.

Crucial to point out here is the fact that the provision not only had primitive tribes as one of the beneficiaries of the act, but also the presence of the provision for the "Other Traditional Forest Dwellers" (OTFD) in this act was intended to benefit "people dependent on the forest for survival." Clause 4(3) of the act defines OTFD as individuals or communities who have "primarily" resided in forest land and have been dependent on it for "at least three generations" "for bona fide livelihood needs" before 13 December 2005 (FRA Amendment Bill, 2014, P2).[2] A generation means a person living in the same area for 25 years. In this regard, sections 4(5), which states that no eviction should take place unless the verification is completed, 4(6), which states that the person should have been in possession of the piece of land on which the individual and or community is occupying, and 6(3), which holds that the duly constituted *gram sabha* must pass the resolution of transfer of land,[3] are important. The question thus arises as to whether these rules were followed while the state government acquired land and diverted forest land to the company for other purposes?

For any acquisition of forest land by any company, the application of the Forest Rights Act is necessary. This was true for the POSCO project as well, for out of 4,004 acres of land assigned to the POSCO 3,096 hectares officially had been declared as forest land by the state government on which other forest dwellers have been dependent for more than the stipulated 75 years. But this provision was systematically violated. The *gram sabha* of Dhinkia had passed the resolution to claim the rights under Other Forest Dweller Rights (Posco Pratirodh Sangram Samiti, communication to Jairam Ramesh, 2008). The *gram sabha* of Dhinkia had held a meeting after three months of promulgation of the FRA. On 23 March 2008, the *gram sabha* had passed a resolution in which the following points were covered: first, it had initiated the process of finding out the rights of the forest dwellers under section 6(1) of the Forest Rights Act; it had started demarcating the boundaries of the village "community forest resources"; and it had resolved to safeguard them. The *gram sabha* had resolved that it would not divert land to POSCO. The forest land if diverted would violate section 5 of the act (Posco Pratirodh Sangram Samiti, communication to Jairam Ramesh, 2008). All these provisions were disrespected. The MoEF, despite submission of the resolution, upheld the illegal forest clearance given to POSCO.

In April 2010, the Environment Ministry and Tribal Affairs Ministry had set up a 20-member team headed by N. C. Saxena to study the factors aiding and/or impeding implementation of the act. Section 6(3) of the act provides that the state government would constitute a subdivisional level committee (SDL) to examine the *gram sabha* resolutions, passed in a duly constituted meeting, to prepare and update the record of forest rights, and subsequently send the files to the district level committee (DLC) – constituted by the state government under section 6(5) of the act – that would take the final decision. N. C. Saxena, the chairperson of a high-powered committee constituted by the Environment Ministry, in his report observed that many SDLCs or DLCs have even been actively discouraging and preventing people to settle the claims in this area (Manthan, 2010: p. 42). The team publicly denounced the lack of progress in recognizing the OTFDs.

A three-member committee constituted by the Ministry of Environment and Forest and headed by Ashish Kothari, the famous environmentalist, visited the POSCO area between April and August 2010 and submitted a comprehensive status report on the Other Traditional Forest Dwellers. The committee, after meeting people and going through the documentary evidence that the movement people placed before the committee, gave the following recommendations:

First, there is a contradiction between the official claim that there was no presence of Other Traditional Forest Dwellers in this area and people's version of holding *palli sabha* to discuss the matter. The district level authorities claimed that since no OTFDs are residing in this area, the state is not bound to consider the consent of *gram sabha* before diverting land to the company. On the contrary, the villagers claimed with evidence that they had the proof that they have been dependent on the forest land since it has been granted by the Bardhaman state a hundred years back. The villagers claimed that the land they are dependent upon was forest land that had been destroyed by the super cyclone of 1999. They were replaced by a new plantation, losing the status of forest land. The committee claimed that the administration knew about this yet did not take the lead to investigate the claim of the villagers. Citing the Ministry of Tribal Affair's letter to all the states dated 9 June 2008 regarding clarifications on the clause 2(c) and 2(o) of the Forest Rights Act, the committee stated that people were dependent on the forest and can be categorized as OTFDs.

The clarification was regarding the use of the word "primarily reside in" and "dependent on" in the previously cited clause 2(c) and 2(o). The circular stated that the word "primarily" mean those individuals "who have habitation" or "patches of land for self-cultivation for livelihood" within the forest area or are working on a patch of land or have a temporary structure, and therefore, spending "primarily" most of their time within the forest area for livelihood, would be considered as Forest Dwelling Tribes and Other Traditional Forest Dwellers (MoTA's letter No.17014/02/2007-PC&V, 2008). The committee, taking cognizance of documents, oral pieces of evidences and clarifications made by the villagers, concluded that there are OTFDS in the area, which is in contrast to the claim made by the district administration (MoEF/MoTA Committee Report, 2010: p. 2).

Second, on the anomaly of holding *palli sabha* and lack of evidence cited by the officials regarding initiation of the process of claimants for the Other Traditional Forest Dwellers, the team concluded that there is discrepancy between the villagers' intimation to district administration about holding the meeting and officials' denial of any acknowledgment of the same. The team highlighted the contradiction at two levels. The district collector mentions holding of the *gram/palli sabha* in a letter number 139 dt 23.02.2010 sent to the MoEF, where it put forward the evidence that there are no eligible OTFD claimants under the FRA. But the collector's office did not attach another letter from the *gram sabha* to the collector, which mentions its resolution not to divert the forest land to the company (MoEF/MoTA Committee Report, 2010: p. 2).

Third, the team found a contradiction in the administration's approach to following one set of rules for one project and another set of rules for the other.

For instance, for setting up of IOCL in a nearby *panchayat*, the administration sought and obtained the *gram sabha*'s consent, while on the other hand, the state and the local administration clearly denied permission to hold a *gram sabha* meeting for the Dhinkia *panchayat* despite several attempts. On procedural grounds, the Kothari report recommended that land acquisition that had started is in contravention to the FRA, and the clearance given to the project on 29 December 2009 breached rules of the FRA. The team thus urged the MoEF to order the Odisha government to halt all kinds of land acquisition until the *palli sabha* gives consent. The reasons for such order are the following: first, the Other Traditional Forest Dwellers are present in this area. Second, there is incomplete identification process of the Other Traditional Forest Dwellers in accordance with the FRA rules. Third is the *gram sabha*'s resolution of refusing to give consent to divert forest land to the project. Finally, on the grounds of deliberate concealment of the fact by the Jagatsinghpur district administration about the resolution of *palli sabha* not to divert forest land for company's use. It also requested the central government to withdraw the forest clearance to POSCO. It also recommended to the MoEF to withdraw clearance to POSCO given on 29 December 2009. It noted, "not doing the above would tantamount to not only ignoring the key objective of the FRA of redressing historical injustice, but also heaping new injustice on the residents of these villages" (MoEF/MoTA Committee Report, 2010: p. 4). Based on the committee's recommendation, the ministry issued a "stop work" order on 5 August 2010, which meant to withhold giving forest land to the POSCO.

The Meena Gupta committee, set up by the UPA government, gave a split verdict on the future of the project. The chairperson, Meena Gupta, gave dissent in the report and recommended POSCO be allowed to set up the factory. The dissent was given on the grounds that the other members were in favor of scrapping the project, because there was no diligent Environmental Impact Assessment carried out in this area, while Meena Gupta was of the view that the project should be permitted to construct the port and that the environmental assessment would be conducted from time to time and suggestions would be given to the project authorities to rectify flaws as they arise out of the studies (Meena Gupta Committee, 2010: p. 5).

Finally, the expiry and renewal of the MoU had called for concern among the villagers. Though the MoU had expired on 22 June 2010, new modalities were chalked out by the central government to renew the MoU after modifying some clauses in it. For instance, the MoU under 19(v) stated that no extension would be granted if the company had not made any "substantial progress" in implementing the project: construction of plant, machinery installation and investment at the site "to the satisfaction of the State Government in these five years as envisaged in the MoU." But despite the existing clause, modalities were discussed to extend the MoU deadline. But it was finally shelved in 2015 because of the nonexistent infrastructure at the project site.

ii Lacunae in Environmental Impact Assessment of the Project

The Environmental Impact Assessment (EIA) has become a tool to protect the environment from industrialization and investors' unsustainable exploitation of resources. EIA as a tool of analysis thus concentrates on conflicts over resources and its constraints. Based on this assumption, the EIA predicts a project's impact not only on the surrounding ecosystem but also on the community as a whole. To that effect, EIA also takes into account participation of the community in the decision-making process. Participation of people in the decision making is a precondition for the democratic federal structure. Even this minimum condition was not followed in assessing the Environmental Impact Assessment in the POSCO area.

Highlighting lapses in the formulation and implementation of development policy was the major task of the movement. The procedural violation in conducting the EIA by the company and the lack of transparency became a substantive issue and bone of contention among three tiers of government *vis-a-vis* the PPSS movement. On 9 August 2007, the Forest Advisory Committee (FAC), which is under the MoEF, took the decision to fell 2.8 lakh trees by POSCO. In the *T. N. Godavaram vs Thurmulpad* case, the Supreme Court had decided that any fresh decision by the FAC on forest clearance should be vetted by the Supreme Court bench. In 2002, the bench constituted the statutory Central Empowered Committee, with wide-ranging power in dealing with impact assessments and hearing applications and passing orders in tune with the Supreme Court's jurisdiction (Manohar and Bhargav, 2016). On 27 April 2007, the Central Empowered Committee regarding the POSCO case had decided that there should not be any piecemeal diversion of land for the project. The impact assessment should take into account the delicate and sensitive ecology of the area and the number of trees felled by the company, and the adequacy and effectiveness of the resettlement and rehabilitation policies should consider a whole array of activities for a comprehensive approach rather than pursuing assessment and seeking clearance for individual projects separately. It should be mentioned here that POSCO in the MoU signed had several components; those included acquiring land for mining, the port area and township area. The Environment Impact Assessment should have been conducted for the project as a whole, not for three projects separately. The Central Empowered Committee (CEC) also ordered that keeping in view the number of trees to be axed, the presence of an NGO is mandatory in the Independent Expert Committee to be constituted by the court that would assess the impact of tree felling and suggest mitigating measures for the area. The CEC recommended "diversion of forest land may be permitted" if the company complied with these norms (CEC Report, 2007). The project was granted final forest clearance on 29 December 2009 despite the previously mentioned statutory provisions. The transgression of statutory provisions under the rule of law is a clear case of misgovernance by all the federative units.

There were procedural lapses in the public hearing as well. The office memorandum of the Ministry of Environment had clarified those project-related public hearings would not take place at the same venue, time and date. The memorandum reflected the spirit of the Delhi High Court order (Writ Petition (civil) No. 9340.2009 and CM Appl nos. 7127/09/2009, Utkarsh Mandal and others). There have been instances in Odisha where separate meetings have been held before. POSCO did not adhere to the procedure. It conducted meetings at one venue and for all thc projects.

The public hearing by POSCO was held 15 kilometers away from the village. This was not a conducive area for the villagers to come, as there was a huge number of police deployed in the area, which intimidated most villagers. Those who had attended complained that their objections were not recorded by the officials conducting the public meeting. This was in complete contrast and contravention to the principles laid down in the EIA notification, 2006. It states that all public consultations should be carried out in a very "systematic, time bound and transparent manner" that should ensure wider public participation. Going by all the procedural aspects, it seemed that there was a complete lack of transparency and procedure for carrying out the project clearance process (for details and the reports see www.newindpress.com/NewsItems.asp?ID=IEQ20070415103516).

Finally, there were inadequate appraisal processes followed by the MoEF-constituted Expert Appraisal Committee (EAC). Three days after Dhinkia *panchayat* conducted the meeting; the Ministry of Environment had received the report from the district and state administration. The expert review committee met for an immediate review of the captive port project. In this meeting they overlooked the letter of the public hearing, took cognizance of POSCO's explanations, dispensed their earlier plan of the site visit and, finally, recommended the clearance of the port project. That rules were side-stepped was evident from the following actions by the government. The government had stated that the objections raised by the villagers would be heard by the expert committee and an Action Taken report would be submitted to the panel of experts within six months of the receiving the letter. The genuine grievances were never addressed by POSCO officials, the state government and the central government. The Action Taken Report was never put before the committee, nor was it available in the public domain through "any kind of suo moto disclosure to the affected people and others who had raised concern" (Letter to Ministry of Environment and Forest by Kanchi Kohli and others, 2010).

Conclusion

The challenge to federalism came from two narratives, one from the company and the other from the PPSS movement against the company and the state. The multinational company was set to take away people's constitutional and traditional rights to land and livelihood by subterfuge, systematically subverting rules and maneuvering the center, state and local governmental institutions. But on the contrary, the movement was countervailing the company's systematic violation of rules and rights through innovative use of the constitutional rules and jurisdictions of different levels of governance that included the court system and a written representation to the government committees, to challenge the unconstitutional method of functioning of three tiers of government. The federal structure rather failed to protect people's rights to decide, act and livelihood. The issue got embroiled in a mesh of power structure. The center, the state and the local level governance worked for POSCO and deliberately scuttled the people's right to decide and act. Even the formal procedures followed by the local bodies following the constitutional provisions were not respected by the federative units at different levels. In this sense, the real issue of empowering people through the devolution and decentralization principle of federalism had no meaning for the villagers. Hence, the movement stressed the need to go beyond the current practice of federalism and adopt innovative ways of responding to people's issues and demands. If federalism wants to move toward democratic consolidation to establish a "federal society" in congruence with the multilevel federal units in India, then it should include within its frame of analysis, as well as in practice, the issues of actors/agency and rights of people to decide, to act and to pursue their livelihood.

Notes

1 According to the NCAER study, the economic benefits of the POSCO would be to the tune of 174,970 crore rupees and would contribute to 33 percent of the net social benefits. For a detailed discussion see NCAER, 2007.

2 Under clause 3(1) in chapter II of the Forest Rights and other Traditional Forest Dwellers Act the government has provided the following rights: "a) right to hold and live in the forest land under the individual or common occupation for habitation or for self-cultivation for livelihood; b) community rights such as nistar by whatever name called, including those used in erstwhile Princely states, Zamindaris or such intermediary regimes; . . . f) Rights in or over disputed lands under any nomenclature in any state where claims are disputed; g) rights for conversion of pattas or leases or grants issued by any local authority or any state government on forest lands to titles."

The act also prescribes that the people retain the rights in 3(1): "h)of settlement and conversion of all forest villages, old habitation, unsurveyed villages and other villages in forests, whether recorded, notified or not into revenue villages; i) right to protect, regenerate or conserve or manage any community forest resource which they have been traditionally protecting and conserving for sustainable use; . . . and finally l) any other traditional right customarily enjoyed by the forest dwelling Scheduled Tribes or other traditional forest dwellers, as the case may be, which are not mentioned in clause a) to k)". Section 2 of the act states that the central government can divert land for small scale development purposes involving one hectare of land and that the "clearance of such developmental projects shall be subject to the condition that the same is recommended by the Gram Sabha."

3 "The state government shall constitute a Sub-Divisional Level Committee to examine the resolution passed by the Gram Sabha and prepare the record of forest rights and forward it through the Sub-Divisional officer to the District Level Committee for a final decision."

References

Central Empowered Committee. 2007. https://stoposco.wordpress.com/2007/11/ (accessed on 02.10.2016).

Chaturvedi, Rohini. 2016. "India's Forest Federalism", *Contemporary South Asia,* 24(1), pp. 1–18,

The Economic Times. 2010. "Government in Overdrive to Launch POSCO Plant". 14 January.

FAO. 2008. *Compulsory Acquisition of Land and Compensation*. Rome: FAO.

Hassel, Anke. 2010. "Multi-Level Governance and Organized Interests", in Henrik Enderlein, Sonja Wälti and Michael Zürn (eds.), *A Handbook on Multi-level Governance*. Cheltenham, UK and Northampton, MA: Edward Elgar, pp. 153–168.

The Hindu. 2016. "Orissa CM Assures of Speeding Up POSCO Work", 26 January.

Hooghe, Liesbet and Gary Marks. 2010. "Types of Multi- level Governance", in Henrik Enderlein, Sonja Wälti and Michael Zürn (eds.), *A Handbook on Multi-level Governance*. Cheltenham, UK and Northampton, MA: Edward Elgar, pp. 17–32.

International Human Rights Clinic, ESCR-Net. 2013. *The Price of Steel: Human Rights and Forced Evictions in the POSCO-India Project*. New York: NYU School of Law.

Letter on Environment and Forest Clearance Related Issues of the Posco Project in Jagatsinghpur: Submission to the Committee Constituted by the Ministry of Environment and Forest on POSCO India Pvt Ltd. Vide orders dated 28th July 2010 and 27th August 2010. https://stoposco.wordpress.com/2010/09/11/submission-to-the-moef-committee-on-posco/

Loughlin, John. 2013. "Reconfiguring the Nation-State: Hybridity vs. Uniformity", in John Loughlin, John Kincaid and Wilfred Swenden (eds.), *Routledge Handbook of Regionalism and Federalism*. London and New York: Routledge, pp. 3–18.

Manohar, P.K. and Praveen Bhargav. 2016. "The Architect of an Omnibus Forest-Protection Case", *The Hindu*.

Manthan. 2010. "Report National Committee on Forest Rights Act", A Joint Committee of Ministry of Environment and Forest and Ministry of Tribal Affairs, New Delhi.

Meena Gupta Committee Report. 2008. Ministry of Environment and Forests, Government of India, New Delhi.

Mining Zone People's Solidarity Group. 2010. "Iron and Steal: The POSCO-India Story", October. http://miningzone.org/

Ministry of Tribal Affairs. 2008. Letter No. 17014/02/2007-PC&V (Vol. II) dated 9th June, 2008, addressed to all States/ UTs, except J&K.

Mishra, Banikant and Birendra Kumar Nayak. 2011. "Paan or POSCO", *Economic and Political Weekly*, XLVI(26 & 27), pp. 12–13.

MoEF/MoTA Committee Report of the Visit to Jagatsinghpur Anti-POSCO and Displacement. 23–24 July 2010. http://environmentportal.in/files/59_MoEF.pdf (accessed on 5.6.2019 at 4.26 pm).

National Center for Advocacy Studies (NCAS). 2010. "Organising and Mobilising in People Centered Advocacy in South Asia: A Resource Book", Pune.
National Council for Applied Economic Research (NCAER). 2007. "Social Cost Benefit Analysis of POSCO steel Project in Orissa", New Delhi.
Noronha, Ligia, Nidhi Srivastava, Divya Datt and P.V. Sridharan. 2009. "Resource Federalism in India: The Case of Minerals", *Economic and Political Weekly*, XLIV(8).
POSCO Pratirodh Sangram Samiti, communication to Jairam Ramesh, Minister of State (independent charge) Environment and Forest. https://stoposco.wordpress.com/page/3/ (accessed on 2 October 2016 at 7.35 pm).
Press Trust of India. 2009. "Posco Gets Final Nod from Environment Min", New Delhi, 31 December.
PUCL. 2011. "Release Abhay Sahoo Now: People's Union for Civil Liberties". https://Kafila.online/2011/12/05
Ramesh, Jairam.2011. Letter to Odisha Government. www.moef.nic.in/downloads/public-information/Posco-final-orders-02052011.pdf (accessed on 19/10/2016 at 1.10 pm).
Roy, Aruna, Nikhil Dey et al. 2013. "Illegal POSCO SEZ", *Economic and Political Weekly*, XLVIII(3), p. 5.
Walti, Sonja. 2010. "Multilevel Environmental Governance", in Henrik Enderlein, Sonja Walti and Michael Zurn (eds.), *A Handbook on Multi-level Governance*. Cheltenham, UK and Northampton, MA: Edward Elgar, pp. 411–422.
Walti, Sonja. 2013. "Federalism and Public Policy: Do Federalism, Regionalism and Hybridity Make Any Difference? Evidence from Environmental Policy", in John Loughlin, John Kincaid and Wifried Swenden (eds.), *Routledge Handbook of Regionalism and Federalism*. New York: Routledge, pp. 116–129.

6

FOREST GOVERNANCE, TRIBAL RIGHTS AND STATE

A study of the third layer of federal structure in India

Kamal Nayan Choubey

The purpose of this chapter is to evaluate the making and working of third layer of federal structure in India's tribal areas, particularly in the context of the enactment and experiences of two laws enacted in the post-liberalization era, that is, the Panchayat (Extension to Schedule Areas) Act 1996 (henceforth PESA) and the Scheduled Tribes and Other Traditional Forest Dwellers (Recognition of Forest Rights) Act 2006 (henceforth FRA). It has been accepted by many scholars of the federal studies that with enactment of the 73rd and 74th amendments to the Indian Constitution in 1993, a "third layer" (Singh and Saxena 2011: 184) or "multilevel federalism" (Arora et al. 2012) has become an integral part of the Indian federal system. Indeed, there are many theoretical discussions related to the asymmetrical federal nature of the Fifth and Sixth Schedule of the Indian Constitution (Saxena 2012; Arora et al. 2012); however, there is no systematic empirical study regarding the impact of the enactment and experiences of two "progressive" laws, that is, PESA and FRA, on the federal structure of India. This chapter enquires as to how far PESA and FRA have been able to create a third layer of federal structure in the Fifth Schedule (in the case of PESA) and other forest areas of the country (in the context of the FRA) and how far they give local communities some "autonomy" and "rights" over forest land and its resources. In this context, it is also pertinent to probe that to what extent this federal structure is functional. Are the rights given by these laws strengthening democracy in these areas, or could they be termed as "tokenism" by the government to resist the Maoist movement in these areas?

The chapter is divided in four sections: the first section describes a brief historical background and key provisions of the centralized or federal characteristic of forest governance in India and the general experiences of both PESA and FRA. The second section focuses on the functional role of the third layer of the federal structure created by PESA and FRA through the case study of the struggle against the Indian Farmers Fertilizers Cooperatives (IFFCO) power plant in the Premnagar Block of the Surajpur district of Chhattisgarh and the use of both these laws by *adivasis* to oppose the acquisition of their farming land and forest land. The third section probes the nature of the third layer of the federal structure, particularly in the context of the idea of asymmetrical federalism, and proposes certain formulations regarding the working of the third layer of federal structure and its overall impact on democratization of the governance of tribal and forest areas. The fourth and final section is the conclusion.

The PESA and the FRA: a framework for the third layer of the federal structure in forest areas

Before colonial rule in India, there was no systematic control of forests by the rulers. They used forest resources according to their need but never tried to control day-to-day activities of forest dwelling communities. The colonial rulers, however, tried to establish their full centralized control over forest land and its resources. For this purpose they established the Forest Department (FD) in 1864 and made laws like the Indian Forest Acts of 1865, 1878 and 1927. The Indian Forest Act of 1878 was based on the idea of "eminent domain," which established that the state was the sovereign of all land under its rule (Singh 1986: 21). In this sense it turned the "customary rights" of forest dwelling communities into "concessions," and those people who could not produce any written document in the support of their claim regarding the ownership of forest land were termed as "encroacher[s]" (Guha 1983a, 1983b; Gadgil and Guha 1992; Rangarajan 1997). However, to quell the resistance of *adivasis*, colonial rulers also made laws like the Scheduled Districts Act, 1874, and created the categories of "Excluded" (for North-Eastern areas) and "Partially Excluded" (tribal areas of rest of the country) areas and tried to ensure certain rights for the tribal population (Bijoy 2012). Indeed, the creation of "excluded" and "partially excluded areas" could be termed as the genesis of an asymmetrical federal structure in these tribal areas, because there were special provisions for these areas that were different from the powers of provincial governments in British India. In most of cases, however, the imperial interests of colonial rulers got priority and the special provisions for these categories were not followed.

In postcolonial India, too, one can identify two parallel processes: Sixth Schedule of the Indian Constitution gives some crucial rights to North-Eastern states – for example, there is a provision in the Constitution for Autonomous District Councils (ADCs). There are some important provisions for Fifth Schedule areas (i.e., tribal areas of rest of India). For example, Section 5 (1) of the Fifth Schedule gives governors of the states with scheduled areas (SAs) powers to refuse the implementation of any laws passed by the Parliament or state legislative assemblies, or he/she can give direction for necessary changes before enforcing them in these areas (see Para 5(1), Constitution of India 2008: 254–255). Obviously, both these constitutional categories are example of substate or region-based constitutional asymmetry in Indian federalism (Saxena 2012).

The Indian government, however, never respected the provisions of the Fifth Schedule, and no governor ever used its power regarding these areas. Dr. B. D. Sharma termed it as "unbroken history of broken promises" (Sharma 2010). Ironically, the Indian Constitution also accepts the validity of the Indian Forest Act of 1927, and the Indian government continuously exploited forests in the name of "national development." This process led to the dispossession of tribal communities from their own forest land and its resources. The Indian government also enhanced its control over the management of forests through legislation like the Wildlife (Protection) Act, 1972, and Forest (Conservation) Act, 1980. Through the 42nd Amendment of the Constitution the "Forest" was transferred from the state list to the concurrent list. All these developments weakened the federal component of the governance of forests in India. Indeed, though there are special provisions in the Indian Constitution for tribal areas, one can easily trace the tendency of centralization regarding the governance of forests in postcolonial India. It worked against the federal potential of certain crucial constitutional provisions, particularly in Fifth Schedule areas.

From the 1970s, however, many grassroots movements emerged in tribal and forest areas against the state-imposed development. The Chipko Movement, started in 1973, and the Narmada Bachao Andolan of 1980s were prominent examples. Maoists also started their activities in forest areas around the 1980s and spread across the forest areas of different states, especially Andhra Pradesh, Jharkhand (then part of Bihar) and Chhattisgarh (then part of Madhya Pradesh). These factors increased the political consciousness of tribals, which led to the demand for the ownership over their land and livelihood resources. I wish to argue that due to a different movement a strong feeling against the centralized management of forest and its resources emerged in hearts of forest dwelling communities, and one can situate the enactment of PESA and FRA in this context.

The PESA was enacted in 1996, and it extends Panchayati Raj institutions in the Fifth Schedule areas. Indeed, its enactment was a result of the demand for more powers by *adivasis* to manage local affairs on the basis of their customs and to ensure their rights over local community lands and resources. It recognizes that the *gram sabha* has the right to be consulted on matters of land acquisition and has the power to act on many crucial issues related to tribal rights in Fifth Schedule areas, such as land alienation, minor forest products and control of money lending to the STs, etc. It underlines that an organic self-governing community, that is, *Tola/para* or hamlet, is the basic unit of self-governance and has the right to preserve their culture and resources (Government of India 1996; Savyasaachi 1998; Choubey 2016). Obviously, it presents a framework for the autonomous *gram sabha* in Schedule V areas.

The FRA, enacted in 2006, was a result of an extensive movement by tribal organizations of the different tribal areas of the country to ensure the rights of forest dwelling communities on forest land and its resources. It gives many rights to forest dwelling STs and Other Traditional Forest Dwellers (OTFDs). It recognizes that every nuclear ST and OTFD family would get "patta" of 4 hectares of "encroached" forest land and makes a provision for the "joint patta" for both husband and wife. The cut of date for STs was demarcated as 13 December 2005, when the Forest Rights Bill was introduced in the Parliament. However, according to the FRA, OTFDs must prove that they have been residing or dependent on a forest land for the last three generations and 75 years before the 13 December 2005. It gives rights to these groups on non-timber forest produces (NTFPs) and the right to manage forests. It accepts the right to habitat and habitation for primitive tribal groups (PTGs). Obviously, it gives STs and OTFDs private and common property rights on forest land and its resources. Also, section 5 of the FRA is related to the duties of the holders of forest rights. It empowers right holders to protect the wildlife, forest and biodiversity (Government of India 2006; Choubey 2013a, 2013b, 2014a, 2015a: 116–180).

It should be noted that the PESA is only for Fifth Schedule areas, but the FRA is for the forest areas of the whole country. The PESA is related to all vital aspects of the village community life of Fifth Schedule areas and gives them autonomy; the FRA, on the other hand, is related to forest land and its resources. Both these laws give important powers to local communities in matters related to land acquisition or the diversion of forest land. Before enactment of the FRA, the process of conversion of forest land for non-forest use was entirely centralized. According to the Forest Conservation Act 1980, the Forest Advisory Committee (FAC) can decide about this, but there is no representation of local communities in it. After enactment of the FRA, the "prior informed consent" of the *gram sabha* became necessary. In this sense both these laws have given crucial rights to the third level of federal structure.

Many studies have underlined that PESA has been partially implemented and there are many instances of their violations (Government of India 2008, 2009; Sharma 1998, 2010; Sundar 2009, 2011; Dandekar and Choudhury 2010: 6–8; Choubey 2014b, 2015b). The following points could be underlined in this context: first, since the *panchayat* is the subject of the state list, every state with the Fifth Schedule areas had to change their *panchayat* acts in accordance with the central PESA. The states, however, only partially implemented it, and some states took many years to make rules for Fifth Schedule areas. For example, Rajasthan passed its state PESA in 1999, but rules were not formulated till 2011. Second, state PESAs have not incorporated all provisions of the central PESA. Of course, this situation led to the subversion of the true spirit of PESA. Third, even in those states where, under the PESA, *gram sabhas* got crucial powers regarding the issue of land acquisition and management of local natural resources, these provisions have been rampantly violated to serve the interests of corporate houses, and one can give many such examples (Choubey 2015b, 2015c).

The implementation of the FRA, too, has not been satisfactory, particularly in the case of community rights. First, although private property rights over the so called "encroached" forest land have been largely implemented, one can find many irregularities in this regard. In many places OTFDs could not get rights for forest land (Choubey 2015a). Second, interestingly, the community rights given by the FRA have largely been unimplemented or partially implemented because they challenge the dominance of the Forest Department. So they created many obstacles in their implementation. Third, though FRA talks about making critical wildlife habitat in national parks and other protected areas so that fewer people would be displaced to create space for animals, the FD totally violated this provision, and in many cases it tried to displace tribals without settling their rights or taking their consent or without any scientific study (Choubey 2015a, 2017).

I, however, wish to propose that there are many areas where local communities organized themselves and used both PESA and FRA to oppose state-imposed "development" projects that could have created problems for their livelihood. Obviously, they have not been successful in all cases. Local tribal communities have also created an autonomous space for themselves through these laws, and in this sense they have strengthened with the idea of an effective third level of federalism. In this context, I will focus on the struggle of tribal communities against the IFFCO plant in the Premnagar Block of the Surjapur district of Chhattisgarh, and also briefly indicate some other examples.

PESA and resistance for livelihood: struggle against the IFFCO power plant in Prem Nagar

The Prem Nagar block is situated in the present Surajpur district of Chhattisgarh. Earlier it was part of Surguja district, but in 2012 a new district, Surajpur, was carved out from it. This area is part of the Fifth Schedule, and due to the provisions of the PESA, STs now have due representation in *panchayats*. In the Prem Nagar block, there was a proposal to establish a power plant, and the local tribal and other people continuously struggled against it. There are unused coal blocks near Prem Nagar, which make it an ideal place for establishing an industrial enterprise.

On 4 June 2005, the Chhattisgarh government and IFFCO signed a memorandum of understanding (MoU). It was decided that IFFCO would set up a 1,000 MW coal-based power plant and mines with the investment of 4,500 crore Rs., and 2,400 hectares of land was needed for the project. It was proposed that 1,100 hectares of forest land, 310 hectares of revenue land and 1,090 hectares of private land would be allotted for this project. Due to this project, the danger of displacement became imminent for at least 31 villages. A mixed population was living in these villages, that is, many STs and lower OBCs castes have been living in these villages for many generations.

A local organization, Margdarshak Seva Sansthan, played a crucial role in making people aware of the dangerous aspects of the proposed project. It organized a meeting of senior citizens/representatives of all the concerned villages in July 2005, and after extensive discussion about the MoU between the government of Chhattisgarh and IFFCO, it became clear to them that it would cause inevitable displacement to many villages. They also realized that even if a village would not be displaced, the proposed power plant would immensely harm their forests and cultural heritage. For the villagers, particularly for the *adivasis*, the forests have not only been source of fulfillment of daily needs, but they are the spaces where their gods and goddesses reside, and almost all rituals, from birth to death, are related to the forests. Since the danger was gigantic, people decided to oppose the project wholeheartedly; they decided to form an organization called Gram Sabha Parishad (GSP), and a resident of Prem Nagar village, Shri Ghashiram, became its president.

Since the PESA has been implemented in Chhattisgarh and elections also are held with due reservation for STs, villagers had some knowledge about this legislation. During my fieldwork in this area, I found that though they did not know all the provisions of PESA related to the powers of the *gram sabha*, they surely understand that it gave rights over their land and resources. Particularly women underlined that constitution (*Samvidhan*) and law had given rights over their land and forests, and the company or government could not snatch away their rights.

In their collective meeting, the senior citizens/representatives of all the villages decided to call the meeting of their respective *gram sabhas* and discuss the issue. As mentioned earlier, there is a provision in the PESA that before land acquisition, the *panchayat* or *gram sabha* should be consulted at the appropriate level. Here, villages called the meeting of *gram sabha* by their own initiative, and out of 31 villages, 27 passed a resolution that they would not give their lands for the proposed power plant. These meetings of *gram sabha* were organized fromOctober 2005 to July 2006. They also decided to start a collective resistance movement against any attempt to acquire their land or the forest land for the power plant.

They decided that they would ensure that IFFCO people could not enter their village and start any work related to the plant. Tribal women played crucial role in this, and many times they compelled IFFCO officials to return back from their village. In the guidance of the MSS and its leader, Mehdilal Yadav, it was decided to resist the proposed project at several level: first, it was decided that all legal mechanisms would be followed to save the land and forests from being acquired. In this context, as mentioned earlier, provisions of PESA were used by the villages. Later, after the notification of the Forest Rights Act, its relevant provisions were also used to ascertain the rights of tribals and other villagers over forests. Second, many measures like *dharna*, demonstrations, hunger strikes, etc., were used to express their resistance to the proposed plant. Third, they used all pertinent constitutional mechanisms to convey their concern about the proposed plant. For example, they sent representation to the president of India, the governor of the Chhattisgarh and the National Commission to Scheduled Tribes, and they also approached MoEF against this project and presented their objection on the basis of its negative impact on biodiversity of the forests.

The local administration tried to break the movement against the plant at many levels. It threatened tribal people who were participating in the movement; it lured many tribals and people from other *adivasis* and non-*adivasis* by offering them jobs; and finally the administration declared Prem Nagar as an urban area. Since there is no separate law for the urban areas of the Fifth Schedule, so the legal provision of discussion with villages before acquisition could easily be avoided. The villagers vociferously opposed this move, and in their unique style of protest they performed all the rituals. Although villagers of Prem Nagar opposed the arbitrary and one-sided decision of the state government, the decision was not changed. Hence, some villagers filed a case against this decision in the Bilaspur High Court, and the final decision of the matter has yet to come.

Finally, due to continuous resistance by the local communities, in 2012 the IFFCO canceled its plant to establish a plant in this area. Indeed, it was a victory for the struggle of tribals and other villagers. It should also be noted that IFFCO has the mining rights in the neighboring Tara Coal Block, so they would preferably search a new place near this area, or it is also possible that after sometime they could again try to revive their plan to establish a power plant in the same area.

Obviously, the people in this area used the PESA and the FRA to save their forests and land. It is, however, also true that they were only part of the many tools used for their struggle. They used nonviolent methods like *dharna* and demonstration but were prepared to go to any extent to save their land and forests. It also created fear and anxiety in the administration, and they could not use any repressive method against the movement. One could say that the struggle of Prem Nagar is an example of how many tribal and lower OBCs in non-tribal communities are using the PESA and the FRA for the resistance of state-imposed "development."

There are many other struggles where local communities used PESA (and also FRA) in their struggle to secure the trio of "Jal, Jangal and Zameen" (water, forest and land). For example, in the anti-Vedanta movement of Odisha, the local activists and Dongaria-Kondh people raised the permission of *gram sabha* before giving clearance to any such projects. Later, when the Government of India constituted the Saxena Committee, it underlined in its report that the provisions of both the PESA and the FRA had been violated in the allocation of land for this project (Government of India 2010a). The Supreme Court in its April 2013 decision made it clear that the decision to give land to the Vedanta Company was a clear violation of PESA and FRA, and the *gram sabha* (*palli sabha*) must decide about the fate of this project. The Odisha State Government, however, called the meeting of only 12 villages because it was easy to manage a fewer number of villages. The state administration left no stone unturned to influence people in the favor of the project (Choubey 2015b: 300–304). But the tribals of all 12villages unequivocally rejected this project. After this, the central government rejected this project in 2014 (Choubey 2014b: 128).

Obviously, in this description, the aim is not present a rosy picture of the implementation of both these laws. As discussed in the previous section, the implementation of both these laws has been dismal. However, the larger point is that at many places local communities are making the third layer of federalism functional, and they are using provisions of these laws to protect their interests and forcing states to respect the structure created by the PESA and the FRA.

Legalism from below: a guiding force for the third layer of federalism in forest areas

As mentioned earlier, there are many provisions for the tribal population in the Fifth Schedule and Sixth Schedule of Indian Constitution. Rekha Saxena has termed these provisions as "Sub-state asymmetries" of Indian federal structure (Saxena 2012: 74).

I argue that both PESA and FRA could be seen as a means to strengthen substate asymmetries of Indian federal structure, and since the FRA is for the forest areas of the whole country, so it creates an asymmetrical federal structure in the whole country. Indeed, since both the laws give some very crucial rights to local communities over forest land and its resources, they have created a third layer of federal structure after the center and state levels. Since both laws have formulated a legal structure regarding land acquisition or changing the nature of forest land, etc., the state machinery has to follow these provisions. In this sense both these laws provide some powerful tools for region-based asymmetrical federalism.

Second, I propose that the idea of asymmetrical federalism could be extended to the implementation of both these laws, too. Some states have performed better than other states in the implementation of these laws, which actually underlines that asymmetrical capacity of local communities to get these laws implemented, even though they are created by India's Parliament and an integral part of the constitutional scheme of the governance of forest areas. We can consider the case of the implementation of the FRA in this context. For example, in Andhra Pradesh, there were 1,69,153 claims related to the individual rights claim till 30 October 2017, and Andhra Pradesh distributed 87,161 titles on the basis of these claims. It simply means that more than 50percent of the claims were not given title till now. In the same period in Gujarat, 1,82,869 persons filed claims for individual rights, and only 81,178 persons were distributed titles, which means that more than 55 percent of claimants were not distributed titles. The case of Bihar is unique; here there were a total of 8,022 claims and only 122 persons were distributed titles (Government of India 2017). The lack of proper resolution of claims could be explained on the basis of absence of strong local level pressure, problems created by the FD and lack of interest from respective state governments. In the absence of individual rights, it is difficult for forest dwelling communities to claim their community rights over forest resources, because they will remain "encroachers" on forest land. Similarly, in the case of PESA, one can clearly underline the difference of function of the third layer of federal structure created by this law. For example, only Chhattisgarh and Madhya Pradesh have made provisions that before acquiring land for development projects, the *gram sabha* will be conducted. In Rajasthan, on the other hand, which had not prepared the rule of the act till 2011, even in the 2011 rules there were no actual powers with *gram sabha* (Choubey 2015c). Obviously, one can underline the asymmetrical functioning of the third layer of federal structure, where the second layer of federal structure created many obstacles in the devolution of actual power to the third layer or created many problems in the actual implementation of the provision of both these laws.

Third, the crucial point regarding the PESA and the FRA is that at many places, tribal communities have developed an understanding of these laws due to their participation in the movement for the enactment and implementation of these laws. In this context, the grassroots tribal organizations have played a significant role. Through village-level meetings they have given tribals an understanding of different provisions of these laws, which led to the emergence of "legalism from below" in tribal areas. It simply means that tribals are now using legal mechanisms to assert their rights over natural resources and to protect their habitat. Indeed, the claim is not that tribal people have developed a full understanding of all aspects of these laws and all tribal communities have developed this kind of understanding regarding them. Many tribal communities, however, have developed knowledge about the relevant provisions of these laws and inculcate the feeling that these laws primarily give them right to live an autonomous life in the domain of the village (in PESA) and also crucial rights to manage forest land and its resources (both PESA and FRA). There are many examples where local communities have started to assert their rights over forestland its resources against the arbitrary control of the FD in different parts of India. Indeed, this "legalism from below" has created awareness among tribal and other forest dwelling communities regarding the important role of both the PESA and the FRA, which has led to the emergence of "insurgent citizenship" in forest areas. Due to this awareness, they have made the third layer of federal structure functional and asserted their rights in many areas. At many places, tribal communities have adopted the spirit of these laws and are using them against state-imposed development projects and to create an autonomous space for themselves. So I propose that in these areas people have accepted the federal sense of decision making. Through their struggles, they have tried to underline that central and state governments cannot decide about their lives in an arbitrary manner.

Fourth, it has been argued that asymmetrical federal features can play a negative role and spread secessionism. The provisions of the PESA and the FRA give actual powers into the hands of tribals and other forest dwelling communities, and there are ample examples that local communities are using them to assert and ensure their rights over their habitat and forest resources. In this sense, the third layer of federal structure created by both these laws has enhanced the faith of tribals in the Constitution of India. The problem, however, is the poor implementation and clear violation of the provisions of these laws and constitutional provisions by the state authorities, which has been providing the Maoists a fertile ground for expanding their support base (Government of India 2008). To overcome these problems, the Indian government should try to implement asymmetrical features related to the rights of tribal communities.

Undoubtedly, there is no scientific tool to measure the level of legitimacy created by both these laws for Indian state and its constitutional scheme. However, since tribal organizations themselves took the initiative for the enactment of both laws and mobilized tribals for them, it could be argued that their enactment had created a legitimacy regarding the constitutional structure and hence strengthened the legitimacy of democracy and democratic struggles. In this sense, asymmetrical features of both these laws are not working for secession from the country or creating a sense of alienation. It should also be noted that these laws should not be treated as "tokenism," since tribal organizations undertook a long struggle for both these laws, and indeed they were successful in making them more "progressive" than their earlier version. For example, the category of OTFD was not part of the FRA when it was presented as a Bill in the Parliament, but later, due to the pressure of the movement, the Joint Parliamentary Committee created this category within the FRA (See, Choubey 2015a). Also, as mentioned earlier, at many places tribal and other forest dwelling communities are using these laws to assert their rights over forestland and its resources.

Conclusion

It is obvious from this discussion that the PESA and the FRA have created a strong third layer of federalism in tribal and forest areas of India. Even though the Indian government has not shown great enthusiasm in giving power to local communities according to the provisions of both acts, there are ample examples of the use of third layer of federal structure by these communities to question the imposition of different "development" projects by the central and state governments. It is, however, also true that these laws have been severely violated in different state and that there has been an asymmetry in the implementation of these laws. The inclination of ruling classes toward neoliberal policies and the influence of corporate houses are primarily responsible for this condition. However, there are certain legal structures that are responsible for undermining both laws. It is a fact that the Indian Forest Act of 1927 is still valid and gives huge powers to the FD regarding the management of forest land and its resources. There are other laws such as the Wild Life (Protection) Act, 1972, and the Forest Conservation Act of 1980 that curtail the rights of forest dwelling communities. This is a situation of "legal pluralism," where different laws are existing and state authorities could easily use one law to overlook another law, which has resulted in the violation of the PESA and the FRA in many areas. However, it should be noted that both these laws have created a mechanism, and they provide a weapon for struggle to the oppressed forest-dwelling communities. Though this third layer of federal structure has many limitations, it should be underlined since they are positive steps toward creation of a vibrant democratic system, which is the essence of the idea of federalism, in the governance of forest areas.

References

Arora, Balveer, K. K. Kailash, Rekha Saxena, and H. Kham Khan Suan. 2012. 'Indian Federalism', in K. C. Suri (ed.), *Political Science, Volume 2: Indian Democracy (ICSSR Research Survey and Explorations, General Editor: Achin Vanaik)*. New Delhi: Oxford University Press, pp. 100–160.

Bijoy, C. R. 2012. *Policy Brief on Panchayat Raj (Extension to Scheduled Areas) Act of 1996*. New Delhi: UNDP.

Choubey, Kamal Nayan. 2013a. 'Do "Pragtisheel" Kanononki Dastan: Rajya', *Jan-Andolanaur Pratirodh, Pratiman: Samay Samaj Sankriti*, 1(1): 149–177.

———. 2013b. *Jangal ka Sangharsh, 'Pragtisheel' Kanoon aur Rajya, Samyik Prakashan, Samajaur Itihas, Naveen Shrinkhla 3*. Delhi: Nehru Samarak Sanghralyaaiyam Pustakalya.

———. 2014a. *The Forest Rights Act and the Politics of Marginal Society*. NMML Occasional Paper: Perspectives in Indian Development, New Series 31. Nehru Memorial Museum and Library.

———. 2014b. *Law as Site of Contestation Between State and the Margin: A Comparative Study of the Experiences of Two Progressive Laws (PESA and FRA)* (Unpublished Final Report of the Post-Doctoral Project Submitted to Nehru Memorial Museum And Library. New Delhi: Teen Murti House.

———. 2015a. *Jungalki Haqdari: Rajneetiaur Sangharsh*. New Delhi: CSDS-Vani Prakashan.

———. 2015b. 'Enhancing PESA: The Unfinished Agenda', *Economic and Political Weekly*, 50(8): 21–23.

———. 2015c. 'The Public Life of a "Progressive" Law: PESA and Gaon Ganarajya (Village Republic)', *Studies in Indian Politics*, 3(2): 247–260.

———. 2016. 'The State, Tribals and Law: The Politics Behind the Enactment of PEA and FRA', *Social Change*, 46(3): 355–370.

———. 2017. 'Turning the Tide in Forest Rights', *Economic and Political Weekly*, 52(1): 21–23.

Constitution of India. 2008. Fourth edition. Allahabad: Central Law Publications.

Dandekar, Ajay and C. Choudhury. 2010. *PESA, Left-wing Extremism and Governance Concerns and Challenges in India's Tribal Districts*, Anand: Institute of Rural Management.

Gadgil, Madhav and Ramachandra Guha. (1992). *This Fissured Land: An Ecological History of India*. New Delhi: Oxford University Press.

Government of India. 1972. *Wild Life (Protection) Act*. New Delhi: Ministry of Law and Justice.

———. 1980. *Forest (Conservation) Act*. New Delhi: Ministry of Law and Justice.

———. 1996. *The Provisions of Panchayats (Extension to the Scheduled Areas) Act, 1996, No 40 of 1996*.

———. 2006. *Report of the Sub Committee Appointed by the Ministry of Panchayati Raj to Draft Model Guidelines to vest Gram Sabhas with Powers as Envisaged in PESA*. New Delhi: Ministry of Pnachayati Raj.

———. 2007. *Scheduled Tribes and Other Traditional Forest Dwellers (Recognition of Forest Rights) Act 2006*. New Delhi: Ministry of Law and Justice.

———. 2008. *Development Challenges in Extremist Affected Areas: Report of an Expert Group to Planning Commission (2008)*. New Delhi: Planning Commission of India.

———. 2009. *Third Report of the Standing Committee on Inter-Sectoral Issues Relating to Tribal Development on Standards of Administration and Governance in Scheduled Areas (Chairman: Dr. Bhalchandra Mungerkar, Member, Planning Commission)*. New Delhi: Ministry of Tribal Affairs.

———. 2010. *Report of the Four Member Committee for Investigation Into the Proposal Submitted to the Orissa Mining Company For Bauxite mining in Niyamgiri*. New Delhi: Ministry of Environment and Forest.

———. 2017. *Status Report on implementation of Scheduled Tribes and Other Tradition Forest Dwellers (Recognition of Forest Rights Act) 2006* [for the period ending at 31.10.2017 Ministry of Tribal Affairs, web address: https://tribal.nic.in/FRA/data/MPROct2017.pdf, retrieved at 02.03.2018.

Guha, Ramachandra. 1983a. 'Forestry in British and Post British India: An Historical Analysis', *Economic and Political Weekly*, 18(44): 1882–1896.

———. 1983b. 'Forestry in British and Post British India: An Historical Analysis', *Economic and Political Weekly*, 18(45/46): 1940–1947.

Holston, J. 2008. *Insurgent Citizenship: Disjunctions of Democracy and Modernity in Brazil*. Princeton: Princeton University Press.

Savyasaachi. 1998. *Tribal Forest-Dwellers and Self-Rule: The Constituent Assembly Debates on the Fifth and Sixth Schedules*. New Delhi: Indian Social Institute.

Saxena, Rekha. 2012. 'Is India a Case of Asymmetrical Federalism?', *Economic and Political Weekly*, XLVII(2): 70–75.

Sharma, B. D. 1998. *The Little Lights in Tiny Mud-pots Defy 50 Years of Anti-'Panchayat' Raj*. New Delhi: Shayog Pustak Kutir.
———. 2010. *Unbroken History of Broken Promises: Indian State and the Tribal People*. New Delhi: Freedom Press and Sahyog Pustak Kuteer.
Singh, Chhatrapati. 1986. *Common Property and Common Poverty: India's Forests, Forest Dwellers and the Law*. New Delhi: Oxford University Press.
Singh, M.P. and Rekha Saxena. 2011. *Indian Politics: Constitutional Foundations and Institutional Functioning*. New Delhi: PHI Learning Private Ltd.
Sundar, Nandini. 2009. 'Framing the Political Imagination: Custom, Democracy and Citizenship', in N. Sundar (ed.), *Legal Grounds: Natural Resources, Identity, and the Law in Jharkhand*. New Delhi: Oxford University Press, pp. 188–215.
———. 2011. 'The Rule of Law and Citizenship in Central India: Post-colonial Dilemmas', *Citizenship Studies*, 15 (3–4): 419–432.
Tartlon, Charles. 1965. 'Symmetry and Asymmetry as Elements of Federalism: A Theoretical Speculation', *Journal of Politics*, 26(4): 873.

7

CLIMATE CHANGE AND SUBNATIONAL GOVERNMENT

Uttar Pradesh

Avantika Singh

This chapter undertakes a microanalysis of Uttar Pradesh's State Action Plan on Climate Change (SAPCC) with the following objectives. Given that the environment is not separately defined as a legislative right under constitutional schedule, and thereby, any subject not listed in part VII of the constitution becomes the center's prerogative owing to its residuary character (Srivastava et al. 2015), the first objective of the study is to empirically probe whether federal complexities relegate and limit the scope of subnational entities to being a mere implementer for enforcing federal policies. Second, how does the state action plan of Uttar Pradesh link with national and international climate initiatives in the broader framework of multilevel governance crisscrossing the horizontal and vertical linkages? Third, the study explores the conceivable drivers of state initiatives or what limits states to be the "laboratories of experiment."

Since decentralization of climate change action plan in India to subnational units is among the colossal exercises being undertaken by countries globally, it offers an academic scope to understand subnational green governance in developing countries, wherein the dominant tendency is to see an environmental issue as a centrally tilted subject.

For testing the aforementioned questions, the study predominately draws from the qualitative research method. A total of 12 semi-structured interviews with bureaucrats, politicians, media, consultants and nongovernmental organizations were included. The questions were exploratory in character to understand the microscopic view of Uttar Pradesh's climate action plan approach in terms of process of design, fiscal requirements, party congruence and center–state relations over the matter. Content analysis was used to critically engage with government reports and to decipher the pattern of communication between federal and subnational unit.

The chapter is broadly divided into two sections. To understand the confluence of climate change and green federalism in India, it is essential to dwell on existing concerns, debates and opportunities. The first section in capturing this dimension will factor in the impact of climate change on the livelihoods of the people. Furthering the theoretical debate, the pendulum oscillating between centralization and decentralization as a guiding principle on environmental matters will be discussed. The last part of this section will approach the issue of climate change with a subsidiarity stance. The second section will empirically focus on the case study based on Uttar Pradesh's SAPCC. Given the complicated scenario, this section, in deciphering the federal component of the climate action plan as formulated in India, will critically comprehend the green federal encounter between the union and the state, the state and the stakeholder and the state and its people to understand the constraints while formulating U.P.'s SAPCC. To ascertain this, three specific dimensions covering an ambit of techno-institutional, sociopolitical and financial dimensions will be examined. Further, U.P.'s SAPCC will be examined on the metrics of "laboratories of experiments." The last leg in the second section will delve into the discussion and conclusive remarks.

Climate change and its potential threat to the livelihood of the people

As a consequence of the neoliberal agenda, the increased anthropogenic activities since the preindustrial era have visibly altered earth's surface temperature across continents, posing a threat to human survival that can further be aggravated by preexisting inequalities. International and national empirical research points at environmental damages emanating from unsustainable activities. The various evident climate risks include the rise in sea level, melting snow caps, delayed winters and increased average surface temperature, etc. According to a report by the World Meteorological Organisation (2017), "the average global temperature from January to September 2017 was approximately 1.1°C above that of the pre-industrial era." Jairam Ramesh, a former environment minister, in a personal communication dated 7 June 2017, shared his fears related to climate change. He defined India as the most vulnerable and exposed to climate change. To him, the "[n]ation's outright dependency on Monsoon, populated coastline, depleting Himalayan iceberg have impacted Indian economic health and vulnerability of its people."

Noting the climate vulnerability of Uttar Pradesh, it is observed that rapid urbanization and the escalated poverty line with an increased incidence of drought and flood makes this densely and overpopulated, vast and agriculturally predominant state a highly sensitive terrain for livelihood. The official reports (U.P.'s SAPCC 2014) anticipate per-year rainfall to increase by 20 percent in the 2050's and an increase in the maximum temperature from 1.8 °C to 2 °C during the same period. This is alarming for small and marginal farmers in the state (accounting for 90 percent of all farmers), as the trend suggests agricultural produce may decline by 25 percent and 50 percent in irrigated and rainfed area, respectively. The situation is deleterious since the 2005 inventory report accorded Uttar Pradesh the status of the most emitting state of India. It contributes about 14 percent of national greenhouse gases (GHG).

Unbalancing the power

Both union and state have an overlapping, concurrent and shared jurisdiction over 52 subjects ranging from forest to factory, as listed in the concurrent list. The state legislature has constitutionally been conferred with full autonomy to deal with subject mentioned in the concurrent list but is subjected to a limitation. If any of the state law is conflicting and contradictory to a union law over a concurrent subject, then the union law will stand, and the state law, to the extent of its volatility, be void. The further prominence of parliamentary legislative power is being reposed in Parliament by specific constitutional articles. To mention a few, Article 249 grants an exclusive right to the parliamentary legislature to legislate over matters of national interest even if they fall under the state list. In case of inconsistency between union and state legislation, the union law shall prevail. The further repose of power with Parliament has been mandated by Article 253, conferring the exclusive right to Parliament in legislating laws conforming to an official environmental stand of India in international politics. In addition to this, Article 51(c) bestows a constitutional duty over the state to adhere to international laws, conventions, treaties and obligations signed by the union. The Biodiversity Act, 2002, legislated by Parliament, was a follow up of obligation agreed upon in the Rio summit, 1992. Further, interpretation by courts maintained and strengthened the union tilt over environmental issues. In *T.N. Godavarman vs Union of India,* for example, the court pronounced that states have to take prior permission from the center before diverting forest land for non-forest use. These changes have tilted the balance of power over environmental issues in favor of Parliament.

Corrective to centrality: is environmental decentralization effective?

Given the over-centralized scenario, academic scholarship tends to suggest that states in India are merely an implementation agency, while the union has a vital role to frame policies. The reasons for abstaining from decentralization as a possible green governance model stationing the overarching power within the union can be deciphered from scholarship critiquing the federal benefits of decentralization. Decentralization scholarship is not free from ambiguity and complexity. Critique doubts its ability to deliver the governance effectively. Apprehension over state and local unit capacity to execute the rule and deliver effective management along with the race to the bottom theory are some of the contentious issues. Therefore, despite the poor performance of a centralized unit of governance, the reason often cited for limited devolutionary power to the decentralized unit has been the concern and doubt over its capacity to carry out the service delivery effectively. Further, race to the bottom theory puts a concern on the role of subnational and local authorities who may purposely lower their regulatory stringency to draw greater financial venture to the region in contrast to other geographical jurisdictions (Fredriksson and Millimet 2002).

In addition, the theory of "interstate externalities" (Revesz 1996) provides an economic rationale for the union federalist tilt in climate change-related policies as opposed to a state tilt. This school of thought rests on the assumption that any small initiative initiating from limited geography bypassing the scope of the international cooperative agreement is economically irrational. In contrast, the work of Kirsten Engel (2006, p. 1022) espouses,

> while not optimal, regulation of climate change by large sub-global governments is not economically irrational. This is because large sub-global governments such as the USA, Japan, the European Union, China are sufficiently large that they can measurably impact the degree of climate change that occurs worldwide.

Therefore, there is an economic driver to cut back emissions. However, Kirsten Engel's analysis is limited in its scope as to this: an economic driver is not present with subnational units for mitigation because of their sheer insignificant size within the union.

Despite these contradictory claims over the capacity of decentralization, Honalde (2001) points out that with an efficacious development of institutional capacity of the decentralized unit as a dynamic process, the capacity at the subnational level may ameliorate given new responsibilities. Further, Jorgensen, too, contradicts the popular assertion about race to the bottom theory and says, "differences in state policy may not necessarily lead to a race to the bottom or exacerbate rivalry. It may even result in positive spill over effects such as drawing lessons from each other" (Jorgensen 2011). In fact, with recent advances, some of the environmental policy processes are categorically conferring their allegiance toward decentralization. India's National Environment Policy (2006) sets forth the foci on the principle of decentralization, that is, "ceding or transfer of power from a central authority to state or local authorities, in order to empower public authorities having jurisdiction at the spatial level at which particular environmental issues are salient, to address these issues."

The argument put forth in favor of decentralization is its efficiency to dodge the limitation of the traditionally centralized governance structure. It has an aptitude to "transfer power and authority, improve efficiency, equity, accountability and inclusion of local people who were previously excluded by the command and control model." (Kiwango et al. 2015).

Through a review of the literature (Broadway et al. 1994; Larson et al. 2004, 2008), the study divides reasons into three main categories to answer why decentralization is perceived to be the best alternative to foster green climate governance.

1 Decentralization leads to increased efficiency of the local unit by way of escalated competition among varied decentralized units.
2 Contrary to the over-centralized toned rhetoric, there is an emerging scholarship accentuating the significance of perceptions of trust, fairness and reciprocity to the effectiveness and efficiency of policy regimes (Ostrom 2010). The feeling of trust is manifested with the decentralized unit's participation. It increases the probability of lowering the cost of climate change-related activities and approximate utilization of the resources.

> Moreover, devolving authority to local jurisdictions engages the interest of all factions at the local level, arguably leading to more healthy local engagement. Accommodating heterogeneity in local tastes allows local jurisdictions to make policy trade-offs that more closely meet local preferences.
>
> (Shobe and Burtraw 2012)

3 Decentralized impetus benefits decision makers with the specific temporal and spatial understanding of natural resources. Further engagement of the "local" helps to substitute any knowledge constraint of policy makers from outside the region on an internalized mechanism. As a corollary, the result is an efficient, inclusive and equity-based sustainable management of the resources with a localized innovation (Larson et al. 2004).

The scholarship culminating from an understanding of decentralization as an enabling tool, and the contradictory rhetoric critiquing the federal benefits of decentralization, helps to strategically situate an issue of concern contributing to the wiser application of the perceived decentralized policy. An analysis thus helps to comprehend some of the real policy choices. The choices are not situated at the loggerhead; rather, the choice is to ascertain what, where and how to decentralize specifics. As Prud'homme (1995) puts in, the problem encountered is not of choice whether the central, regional or local government should be in charge of the devolution of governance over the matter; rather, it is "to organise the joint production of the services by the various levels." When comprehensively crafted and executed, they have the potential to improve public sector efficiency at least in some specific issues. Prud'homme (1995) likens decentralization to a drug: "when prescribed for the relevant illness, at the appropriate moment and in the correct dose, they can have the desired salutary effect; but in the wrong circumstances, they can harm rather than heal."

Climate change subsidiarity principle

The chapter till now has critically engaged with the question of why decentralization should be a guiding principle in natural resource management impacted by climate change. What also is important for our consideration is to analyze factors as to why the state, vested with enormous power and authority, should pursue climate decentralization? Therefore, green decentralization along with an issue of administrative efficiency is also a matter of politics.

The chapter argues that the unitary design in matters of environmental concerns will be limited in achieving its goal. Since India is signatory to the Paris Agreement, its economy will deal in the management of climate concerns on a cost-effective ratio. Devolution of climate change policies to the subnational level guided by a "top to bottom approach" will lack cost effectiveness, and as a subsequent prelude, the unitary model will not achieve the standard cost-effective layout even at the national level. The chapter thus delves into possible arguments reasoning for the greater stake of cooperative federalism, which unlike the past precedence of a central tilt adopts the greater cohesion and symmetry in the assigned federal roles based on the principle of subsidiarity. Climate change with its spillover impact has the potential to change this status quo, as the federal jurisdiction cannot be contained within geographical limits. Hence, it cannot be regulated by apportioning the responsibility to a particular governance level. Green federalism with the subsidiarity narrative acknowledges this flexibility and leaves the authority to the subnational and local levels to manage the deficit unless there is a compelling reason as to why the power should rest with the union (Golub 1996). It is to be noted that distribution of power has not been regulated under its scope; rather, it deciphers the way out in governing and justifying the use of such powers as per specific case (Lenaerts 1993).

The case study of Uttar Pradesh's SAPCC

A countrywide assessment on the impact of climate change was done as reciprocation to global level monitoring and assessment (Press Information Bureau, GoI 2009). Sequentially, a National Action Plan on Climate Change (NAPCC) (GoI 2008), as a result of the impending requirement to draft a nationwide climate plan, was formulated. It is an overview of the country's readiness to withstand climate-related stress. The plan highlights sustainable development as a guiding principle to deal with climate change issues. The core of this plan – the eight national missions – are providing a multipronged approach to achieve critical goals while addressing climate change.

NAPCC's document has been in the public forum since 2008, after which an advisory was issued to the states to formulate their individual SAPCC by 2011 in line with the eight missions of NAPCC. Financial assistance was extended by the Ministry of Environment and Forest (MoEF) and asked the state to outsource services of agencies, approach donor agencies and consult expert commentary if need be. The purpose was to decentralize the actions beyond the union purview and to develop adaptation interventions for state subjects such as agriculture and water. State plans, once finalized, were put under the scrutiny of an expert committee on climate change and the national steering committee on climate change, with officials from the MoEF and relevant central ministries. The expert committee used NAPCC as a central reference guiding principle in endorsing state-based climate action plans (MoEF 2015).

Following the central mandate, the Department of Environment of the Government of Uttar Pradesh released its final draft version in 2014, mentioning seven sectorial missions akin to NAPCC. Some components of the sustainable Himalayan mission have been as of now secured under the strategic knowledge mission, and later a detailed mission document is to be set up in deliberation with different states in the Himalayan region in order to have an active and coordinated arrangement that extends beyond the state (U.P.'s SAPCC 2014). The overall objective of the plan is, "to define state adaptation and mitigation priorities in synchronisation to the national action plan on climate change with sustainable development as its goal" (The principal secretary environment, Government of Uttar Pradesh, personal communication, 27 September 2016).

The course of action in preparation of the SAPCC has been recognized in the methods mentioned in Figure 7.1.

Analyzing the structure and objective of NAPCC and SAPCC (Uttar Pradesh) on the federal plank of multilevel governance, which by definition covers the horizontal and vertical linkages between governments and various actors across spatial and sectorial scales (Corfee-Morlot et al. 2009), it can be deduced that U.P.'s SAPCC is engrafted in the multilevel governance of climate change. It bridges the crucial policy gap between the international, national and subnational units as Uttar Pradesh SAPCC's suffice with the obligation and decision taken at the international climate change forum, en route to its compliance with the centrally mandated climate document, NAPCC. In fact, as the Government of India has put in, a viable recommendation in SAPCC may be incorporated in India's official statement to combat climate change. The U.P., in direct capacity, has also used as a resource the expertise of an international and national agency – GIZ and CTRAN

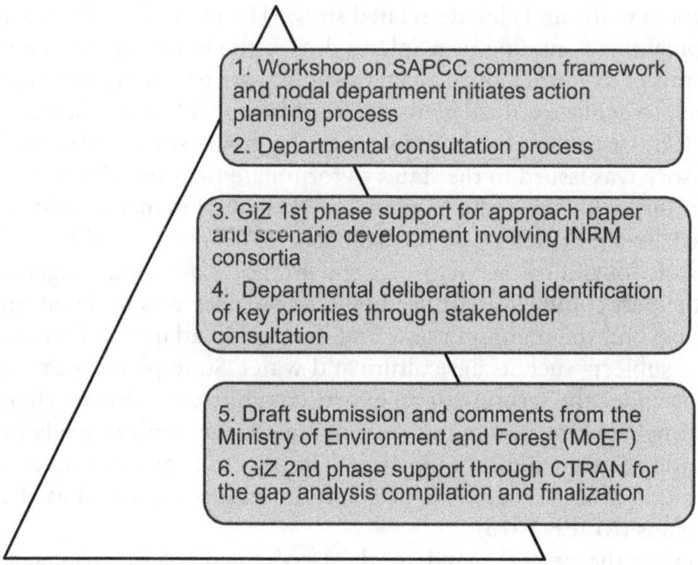

Figure 7.1 Key Approach Followed for Preparation of U.P. SAPCC
Source: State Action Plan on Climate Change, Department of Environment, Government of Uttar Pradesh (2014)

Consulting. The initial workshop to discuss the process and component of the climate draft involved substantial stakeholder deliberation. Thus, at an upfront institutional level, the federal component box can be ticked off. However, a microscopic detailing between the layers of deliberation in the multi-governance framework provides a different picture, embarking on percolation of a faulty decentralization mechanism. For example, at the federal level, NAPCC's mainstreaming of objectives and its state-wide implementation signifies a hierarchal modus operandi. While national goals were formulated by the union, the obligation of implementation rest with the subnational unit. In fact, NAPCC lacked an integrated, inclusive approach while formulating the national document; it did not invite inputs from lower subnational ends (in a bottom-up approach). Interestingly, however, MoEF entrusted itself the task to review the draft submission of SAPCC's by states on the basis of a common framework provided to it by MoEF (top-down approach). Another stark example of this problematic can be cited on SAPCC's structure, whereby a nodal agency is the only source entrusted with the task of giving suggestive policy actions to various departments, and it is left to the wisdom of these departments to adhere to the suggestions or not. Therefore, strategies lack interconnectedness with other sectoral strategies.

Given the complex scenario, this section, about deciphering the federal component of the climate action plan as formulated in India, will critically comprehend the green federal encounter between the union and the state, the state and stakeholders, the state and the people to understand the constraints while formulating U.P.'s SAPCC. To ascertain the same, three specific dimensions covering an ambit of techno-institutional, sociopolitical and financial dimensions is put under scrutiny.

Understanding the problematic: three-way approach

The techno – institutional dimension

U.P. took a reasonably long time to conceive the draft plan and missed critical local engagement

The center fixed 2011 as the deadline to submit the draft document, and U.P.'s final draft was in public forum in 2014. A fairly senior bureaucrat, in response to a question – "What caused the delay in U.P.'s draft?" – said, "financial constraint to take up the whole exercise of documenting proofs, giving payments to external agencies, etc. can be attributed as a reason," (personal communication, 28 September 2016). Further, the essence assessed from all the stakeholders interviewed in Uttar Pradesh on the possible causes which led to such a delay was that climate change is not a high profile and vote bank issue in a state marred by caste and class politics. NAPCC was also not serious regarding providing timely assistance and training in formulating the climate change draft.

Already behind time, the completion of the draft was rather a hurriedly done exercise, as

> in the midst of many limitations in terms of accessing information, know how on how to draft the draft, lest it was important to complete the draft with pressure from the center. Thus whatever has been missed like vulnerability assessment will be now concluded.
> (personal communication, 26 September 2016)

Therefore, the bureaucratic energy was spent on completion of the draft. Being late, the documentation of climate change vulnerability in a local context to frame policy measures was missed.

U.P.'s SAPCC is more of a sustainable document than a climate change issue

Uttar Pradesh, in all conformity, have understood the climate change document as a default sustainable development program that may use the climate lens as per its convenience. The reason factoring to such gap accrues to the state's underdeveloped capacity to initiate convenience in comprehending the state's climate reality. As noted from bureaucratic and stakeholders interviews (personal communications 25– 28 September 2016), state documentation of climate vulnerability was done amid a knowledge deficit. This resulted in the absence of holistic planning to identify and prioritize activities keeping in purview the climate reality and demands at the core. Hence, there is considerable overlap between the existing development policies and drafted climate resilience proposal.

Adding to the observation made by Jogesh and Dubash (2015) in context of Madhya Pradesh SAPCC, U.P.'s SAPCC, too, with an overarching focus on adaptation limits itself to state development projects and is not tending to long-term climate resilience. Additional action on climate change is not being identified.

U.P.'s SAPCC is being driven by an impulse of adaptation

Even if decarbonization of the economy happens today, it will take decades to achieve a 2 °C stabilization pathway, as agreed by all members of the United Nations Framework Convention on Climate Change (UNFCCC). Therefore, adaptation efforts that attempt to increase people's resilience capacity must constitute a vital mechanism to combat the impact of climate change already locked in by past and current emission rates.

In the context of India, its domestic action plan on climate change is primarily driven by a discourse on equity that in global negotiation has transpired from "historical responsibility" and the "polluter pay principle," which non-Annex I countries are not the part of. Consequently, its focus is centered on adaptation and is limited in its scope on mitigation. Further devolution of climate redressal mechanisms to the subnational unit, as Jairam Ramesh, former minister of environment and forests, stated (personal communication, 7th June 2017), was to evolve state-level adaptation action plans.

U.P.'s SAPCC, thus, in conformity with the federal suggestion, evolved the final draft document, with a heavy tilt toward adaptation measures. U.P.'s budgetary allocation, as mentioned in SAPCC, apportions approximately 70 percent of the total estimated climate budget to adaptation-centric measures.

The problem with the U.P.'s adaptation-centric draft emerges from three sources; First, adaptation efforts are broadly positioned within an ambit of the developmental paradigm, reducing the scope of adaptation to procedurally deal with the socioeconomic deficit and advance for economic growth. Amalgamation of the two divergent yet occasionally overlapping policy issues creates a myth. Development is not essentially adaptation, as it may not be concerned with building the resilience of the community. Second, while the sectorial identification along the adaptation strategies has been identified, it missed the mainstreaming detail. Third, the center's inherent confusion further adds to the problematic. At the outset, states were asked by the union to align their draft with NAPCC'S sectorial objectives, which also constitutes mitigations measures such as solar and enhanced energy missions. Further things were muddled by the informal request by the center to not include a greenhouse gas (GHG) inventory, apart from what was mentioned in NAPCC. Despite that, since the request was not bureaucratically channelized, states like Uttar Pradesh have in their climate framework included a section on their GHG inventory.

Local vulnerability assessment

A missed opportunity: in an interview conducted with an NGO activist, a participant in drafting the SAPCC document said, "The climate change draft has not approximately incorporated the suggestions thrown by practitioners. It seemed they are more focused to design a draft in line with NAPCC rather than the following the bottom up strategy" (personal communication, 27 September 2016). Thus, the SAPCC has missed by choice the local vulnerability context, and by the time of the drafting of the SAPCC, the recommendations were attuned to the broader strategy rather than place-based vulnerability.

Further, adding to the difficulty, an official interviewed during the fieldwork suggested that, "with lack of technical support and training, U.P. is virtually handicap as there exists no clarity to guide them in processing with micro-level and sector wise vulnerability assessment" (personal communication, 26 September 2016). The reason for this failure is the result of not effectively devolving authority to an organization working on the ground in purview to mainstream their expertise. It is evident that groups with preexisting vulnerable characteristics, geography, etc., have not been sectorally identified and prioritized.

Imported models used to depict regional projections

A common framework of SAPCC being circulated by MoEF expects states to present their climate change scenario and projections at a relevant spatial and temporal scale. Uttar Pradesh, which laid its report in 2014, in response to the common framework requirement, used the Indian Meteorological Department (IMD) gridded rainfall/temperature data from 1971–2005 and 1969–2005. Further, for physical climate projection, it imported climate modelling based on IPCC PRECIS; for the socioeconomic scenario it imported the A1B IPCC SRES model and Q14 QUMP model for simulation. "These global models have their limitations to simulate the finer regional features and changes in the climate arising over sub-seasonal and smaller spatial scales" (Kumar 2018). Given the uncertain nature of the monsoon being driven by unique physiographic features, this becomes more relevant to the Indian case.

Sociopolitical dimension

Party congruence

As a senior bureaucrat put it, "SAPCC is a process which will culminate along its journey. It has no impact on the change in government and does not impact the conditions leading to change in government. It is apolitical in character" (personal communication, 25 September 2016). Drawing an assertion from Katzenstein et al. (2001) work on social movements, further triangulated by interviews with bureaucrats, the study notes that climate change in U.P. is a subject of "interest politics" since it lacks electoral card sheen. Party manifestos of all the major parties are drought ridden in terms of factoring in even the term "climate change." To add to the apathy, "there is sparse or to be precise no media coverage on climate change related issues/concerns/initiatives" (Sharat Pradhan, personal communication, 27 September 2016). It thus gets encapsulated within the framework of the bureaucratic maze and court orders. We further argue, based on another research done across a few villages in Uttar Pradesh (Singh 2017), that for 80 percent of the interviewed respondents, climate change is not seen as a serious voting parameter; however, the majority of respondents confirmed the negative impact of climate change on their livelihood. The nonseriousness is the result of two things: first, unlike the green politics in European countries, which also is now an emerging trend in US politics, India does not have dedicated green parties. Further, the people in Uttar Pradesh primarily engage with politics on issues of identities; thereby, political parties reflect on those divides. Second, many states in India, along with Uttar Pradesh, are disinterested in exploring SAPCC with a serious concern to be laboratories of change, because the cost of climate-related activities is higher than the benefits gained from them. Since states in India are constitutionally bounded with an accorded duty 51(c) to adhere to the union's international stand, they are taking the formation of SAPCC as business as usual.

SAPCC does not reflect people's sensitivity

U.P.'s framework, which is reflective of existing rhetoric on sustainable development as a default concern, and thus limited in its vision, fails to examine implications that climate change can bring about on sustainable life choices and thereafter impact future development initiatives. Though SAPCC should be well credited to bring in sustainable environmental concerns and priorities into exclusive policy focus, the document, however, fails to put in place the long-term concerns of climate change. The document strictly works in sectorial priorities and assumes that its benefits will be percolated to the people automatically. Human beings are not the center recipients of the hypothesized policy benefits. SAPCC fails to understand that power, caste, gender and class-ridden societies like U.P. do not automatically distribute the equitable benefits, and those vulnerable in negotiations due to their preexisting vulnerability are further socially excluded by not being able to receive the perceived benefits. The technical gloss is deaf and blind in locating people's specific concerns and vulnerability in the context of climate change. Akin to the NAPCC, U.P.'s SAPCC, too, mirrors its parent document, posting a word of concern but without giving any effective recourse.

In the given purview, it is not strange to notice an initial skepticism among the bureaucratic cadre about the link between these social realities and climate change. "Government is taking pro-active actions on issues like gender and caste to empower them. You can dig in more detail from the respective department" (personal communication, 25 September 2016). There is thus a tendency to see the social factors as a separate category than climate change, which serves no direct linkages.

An extent of the stakeholder consultation to formulate SAPCC

The common framework document designated for the state's guidance mentions the importance of the collaborative role of stakeholders as "maximizing the perspectives and to increase the robustness of analysis" while "ensuring its quality, consistency, relevance, pertinence, and transparency." However, no further guideline or proper methodology for stakeholder's consultation was provided to states by the Ministry of Environment, Forest and Climate Change (MoEFCC).

In the case of Uttar Pradesh, while the SAPCC preparation process reflects stakeholders' interventions, making the process open and integrative, the citation of their stakes have not necessarily been worded in the document. Two cardinal reasons, as deduced from stakeholders interviews, are first, the state's reluctance to imbibe something that is out of the box not in tandem with union document, and second, the technical nature of SAPCC made it redundant to incorporate the local social perspective as a vantage point to craft the policy.

Financial dimension

Budgetary issues in SAPCC document

Mapping the budgetary character of the U.P., the estimation to implement SAPCC in the U.P. for 2014–2018 is Rs. 46,946 crores in 93 actions that include both adaptation and mitigation. However, this budget is preliminary. Sectorial mission and nodal departments are expected to update it periodically. The U.P. additionally draws from an already initiated state's development project and centrally funded schemes to meet its budgetary requirements.

A close look at the proposed budgetary allocation (Table 7.1) highlights that the maximum leverage has been given to the water sector (Jal Mission), sustainable habitat and Green U.P. Mission. While the maximum budgetary allocation in the vulnerable water sector is justified, it is astounding to notice the 0.2 percent of the meager budget allocation apportioned to the agriculture sector, on which approximately 70 percent of the population is dependent for its livelihood. It is a well-recorded fact that due to the failed monsoon, increased flood and drought incidences impacting crop production; farmers across U.P. have suffered substantial economic loss. Thus, the hypothesized objective of the multidimensional budget to "improve human conditions and capabilities, enhancing the standard of living for poor and marginalised groups to prepare themselves to cope with adverse impacts of climate change" (U.P.'s SAPCC 2014), falls fatally. With such alarming data available, not incorporating their vulnerability with a higher budget allocation shows poor management and targeting capacity by the SAPCC. Further, an overarching tilt of state toward urban centers can be registered, with 28 percent of the budget being allocated to the sustainable habitat mission (Kumar 2018).

Despite the vulnerability of this vast state with a dense population and constrained by its financial ability, the union has only financed two major demonstration projects for this U.P. along with other states from its National Adaptation Fund, which functions under MoEFCC. First is an Ecosystem services-based adaptation to climate change in the Bundelkhand region through the Forest and Wildlife Department, Government of Uttar Pradesh, and second, developing climate resilience in rural areas through crop residue management through the Department of Agriculture, Government of Uttar Pradesh.

Table 7.1 Financial Budget for SAPCC – U.P. (2014–2018)

Mission	*Adaptation*	*Mitigation*	*Both*	*Total*	*No. of Priorities*
Sustainable Agriculture Mission	74.75	21.50	6.50	102.75	11
Solar Mission	..	449.75	..	449.75	5
Energy Efficiency Mission	..	..	..	..	11
Green U.P. Mission	4,469.85	..	4,010.65	8,480.50	7
Strategic Knowledge Mission	136.00	..	..	136.00	16
Jal Mission	24,175.77	224.90	..	24,400.67	27
Sustainable Habitat Mission	1,524.53	11,851.37	..	13,375.90	16
TOTAL	30,380.89	12,547.52	4,017.15	46,945.56	93

Source: State Action Plan on Climate Change, Department of Environment, Government of Uttar Pradesh (2014).

U.P. SAPCC: lack of an innovation to be a lab of experiment

Laboratories of experimentation provide a scope to introduce new ideas, innovations, and initiatives, different from the existing, while processing a policy design and outcome (Jordan and Lenschow 2008). The subnational unit in US politics exhibits that they have potential to act as a laboratory of experimentation by being open to new ideas and innovations and thus allowing room to be a potential testing ground. Reviewing the same in the context of Uttar Pradesh climate document, this chapter discovers the existence of some grave uncertainties, observing that conditions are still not ripe in asserting the U.P. SAPCC as a laboratory of experimentation because of its lack of lustrous performance in innovation and inciting new ideas and initiatives. The genesis of the initiated process to draft a climate framework was not an attempt spurred by consciousness of the problem; instead, the exercise was undertaken because of the union mandate. In fact, efforts were made so that SAPCC assumes a character of being an iterative reflection of its parent document. In an interview with a top-level bureaucrat (personal communication 27 September 2016) in Uttar Pradesh, it was observed that along with an inherent administrative laziness, state officials feared that nonconformity of SAPCC with NAPCC would lead to a further bureaucratic process to amend the provisions in conformity with the union plan. This adherence clause with the parent document, NAPCC, puts to the question the scope available with states to really craft a document and be a laboratory of experiments in the real sense.

Why U.P. is underperforming – a case of the renewable energy sector in India

Indian states like Tamil Nadu, Karnataka, Gujarat, Rajasthan are working exceptionally well in the field of renewable energy driven by the economic/political agenda, even crossing beyond the set national objective as mentioned in NAPCC. For example, Tamil Nadu's share in the wind energy sector accounts for 43.94 percent as per 2011 data. The analysis of this success rests in subnational jurisdiction over climate-related sectors like renewable energy, which offers an economic impulse to states to not only contribute to India's set standard in meeting the global demand but also an opportunity to indulge in green business. This tendency has led specific state initiatives to outperform other states and the federal benchmark. This implication can directly be contrasted with the work of Rabe et al. (2006), which analyzed the US context to understand why certain single state units in the US may altogether ignore the collective climate action problem and work independently of state or federal unit. To him, climate policies by the subnational unit in the US are driven by, "co-benefits of climate mitigation, the competitive assets of the states and their interest in maximizing their economic stability and welfare" (Rabe et al. 2006). He further adds, "states that own these assets will push and try to upload their mitigation strategies to the higher level" (ibid, p. 44). This bottom-up approach thus overcomes the federal financial constraint while also advocating that policy measures in a country as vast as India, with a multilevel climate governance structure, cannot be solicited only through a top-down approach (Jorgensen 2011).

However, in the context of the Uttar Pradesh climate action program, the challenges are manifold. As concluded from stakeholders' interviews (personal communication 26th September 2016), among the reasons as to what explains the relative underperformance of U.P. as compared to other states, who are able to outperform federally set objective on climate change, the conclusion was U.P.'s outright focus on adaptation measures accounting to 70 percent of the total estimated climate budget. This heavy budget tilt and focus on being in tandem with the federal directive, according to stakeholders, takes away an opportunity to invest in mitigation activities that have a scope to attract green economic investment/carbon credit business. It is worth mentioning that the opportunity to cut back on toxic gases is enormous, as U.P. is among the highest carbon emitters. However, further research is required to what seems to be a simplistic response in excusing the state's unresponsiveness.

Discussion

The study postulates four theoretical premises to conclude its analytical discussion:

1 While climate change brings uncertainties and complexities to the nation's life, climate change measures in the green federalism discourse offer initiatives of unique character. This fosters regional agreements and alliance across the nation. The study thus adds to the scholarship of the race to the bottom theory, seeing climate change as an opportunity allowing state and local governments to engage in competition with their peers, where they assume the responsibility to make their best strategic responses to national initiatives. Therefore, the state assumes an essential role of being executors, innovators and initiators in climate policy and environmental decision making.
2 Constitutionally, the national government in India is not mandated to take an overarching charge of all the environmental-related subjects except in some specific mentioned constitutional provisions. In an event of changing the federal structural characteristics, there will be a political cost involved in establishing such a policy regime. In such a scenario, climate change thus offers an ideal case for cooperative federalism, where the national strategy on climate change, for its successful execution, requires that the policy scale down to all levels of governance and

streams through the programs and projects of all relevant ministries/departments and stakeholders' actions. The issue of climate change further mandates an all-encompassing green federal approach subsumed in the subsidiarity narrative. The principle of subsidiarity discards the scholarship that enforces the prime responsibility to the national government in setting forth the climate change blueprint. It offers the devolution of power for managing the deficit by repositioning the authority at the best federal location where its administration will result in maximum advantage to the people.

3 The chapter is in coherence with Kirsten Engel's work (2006), which postulates that curbing the emission rate by individual countries is economically viable, but it also puts a challenge to her rationale, limiting its scope by discarding the role of the subnational governments to be economically rational because of their size and absence of an economic driver to mitigate. As a critique to the latter response by Engel, this chapter asserts that since the economy at the subnational level together forms the resource base of a country, the emissions discharged from the resources thus accounts to the total production of greenhouse gases. Therefore, any contribution at the subnational level significantly adds to the national mandate, and this very proposition makes it economically viable. The study, further drawing from Rabe et al. (2006), explains that states in a multi-layered federal structure who possess the competitive assets try to push the mitigative activities to maximize their economic stability and welfare. Therefore, it is not, per se, only the division of authority dispersed across governance units determining which national policy can and ought to be implemented. The incentive garnered and distributed within the federal system will impact the effectiveness of contributions by all levels of governance.

4 Further, the chapter asserts that climate change policy that is crafted on an inherent assumption of harmonization of subnational policy around an optimal national policy is not viable and in fact may inundate the implementation of national policy as futile and inefficacious. The unlikely culmination of cost efficiency/benefit may fail to induce subnational governments to take over nationally defined targets in cutting down the greenhouse gas emission.

Conclusion

Different systems have dealt with the environment and its domains differently depending on the structure of government and the stage of development and environmental governance. The U.P.'s sectorial categorization mirrors the NAPCC sectorial bifurcation, but within that, the U.P.'s SAPCC was able to infuse the local conditionality flavor. Local data were statistically processed, and consequently, an action plan reflective of this was designed along with financial requirements. The chapter reflects that the process of climate change framework in Uttar Pradesh worked through a multilevel governance structure, though in a limited way. The horizontal linkages in terms of the state's arm extension to agencies and stakeholders have brought in new insight and additional value by channelizing the conversation. However, the vertical linkage between national and subnational units in the domain of multi-governance failed to excite the new innovations and have somewhat contributed to the limitation on the authority of the subnational unit to groom itself as a laboratory of experiments. The input in terms of what goes into the making of climate policy was heavily dictated by a bureaucratic modus operandi, and extreme caution was taken to reflect the center's dictate. As a result, the linkage has failed to foster the long-term resilience capacity of the state or out-of-the-box sectorial suggestions. As a result, the policy output was propelled to intrigue development rather than reflecting long-term climate resilience concerns offering considerable policy actions of an amplitude compounding India's contribution internationally.

The chapter concludes that the state's exercise was not notably a consciousness-driven initiative, and as is clear from its problematic budgetary allocation, it is not reflective of the local vulnerability context. Despite the limitation of U.P.'s SAPCC expressed by the chapter, the future is expected to take a climate-sensitive turn in Uttar Pradesh. This inference is based on the embryonic development, as per reports in 2017, that Uttar Pradesh is set to institute India's first independent climate change authority. This legal body will work under the leadership of the chief minister to actualize the seven sectorial missions mentioned in SAPCC. The department is set to work in their capacity, and the newly constituted climate change authority will coordinate the activities. Uttar Pradesh was also set to contribute 1,000 MW of solar energy in 2017–18, thus internationally contributing to India's self-commitment under the Paris climate change accord to build a 60 GW solar project by 2022 (*Business Standard*, February 8, 2018). This policy measure exemplifies the novel retreat from the status quo.

References

Broadway, R. et al. 1994. "Fiscal Federalism Dimensions of Tax Reform', *Working Paper*, pp. 3–85. Washington: The World Bank.

Corfee-Morlot, J. et al. 2009. 'Cities, Climate Change and Multilevel Governance', *OECD Environmental Working Papers 14*. OECD.

Engel, Kirsten H. 2006. 'State and Local Climate Initiatives: What Is Motivating State and Local Governments to Address a Global Problem and What Does This Say about Federalism and Environmental Law?', *The Urban Lawyer*, 38: 1015–1029. Arizona Legal Studies Discussion Paper No. 06–36.

Fredrinksson, P.G. and D.L. Millimet. 2002. 'Strategic Interaction and the Determination of Environmental Policy Across US States', *Journal of Urban Economics*, 51: 101–122.

Golub, J. 1996. 'Sovereignty and Subsidiarity in EU Environmental Policy', *Political Studies*, 44(1): 686–703.

Government of India. 2008. *National Action Plan on Climate Change*. http://pmindia.nic.in/Pg01-52.pdf (accessed on 02/06/2018).

Government of Uttar Pradesh. 2014. *State Action Plan on Climate Change, Department of Environment*. Uttar Pradesh: Department of Environment.

Honalde, B.W. 2001. 'Theoretical and Practical Issues of Local Government Capacity in an Era of Devolution', *The Journal of Regional Analysis and Policy*, 31(1): 78–90.

'India's Renewable Energy Project Tendering to Be an All-time High in Fy18', *Business Standard*, February 2, 2018. New Delhi (accessed on April 4, 2018).

Jogesh, Anu and Navroz K. Dubash. 2015. 'State-led Experimentation or Centrally-motivated Replication? A Study of State Action Plans on Climate Change in India', *Journal of Integrative Environmental Sciences*, 12(4): 247–266.

Jordan, A. and Andrea Lenschow. 2008. 'Integrating the Environment for Sustainable Development: An Introduction', in A. Jordan and Andrea Lenschow (eds.), *Innovation in Environmental Policy? Integrating the Environment for Sustainable Development*, pp. 3–23. Cheltenham: Edward Elgar.

Jorgensen, K. 2011. 'Climate Initiatives at the Subnational Level of the Indian States and Their Interplay with Federal Policies', *Montreal: Paper Presented at the 2011 ISA Annual Convention*, 16–19 March 2011.

Katzenstein, M. et al. 2001. 'Social Movement Politics in India: Institutions, Interest and Identities', in Atul Kohli (ed.), *The Success of India's Democracy*, pp. 242–269. New York: Cambridge University Press.

Kiwango, W.A. et al. 2015. 'Decentralised Environmental Governance: A Refection on Its Role in Shaping Wildlife Management Areas in Tanzania', *Tropical Conservation Sciences*, 8(4): 1080–1097.

Kumar, Vineet. 2018. *Coping with Climate Change: An Analysis of India's State Action Plans on Climate Change*. New Delhi: Centre for Science and Environment.

Larson, A.M. and J. Ribot. 2004. 'Democratic Decentralisation Through a Natural Resource Lens: An Introduction', *The European Journal of Development Research*, 16(1): 1–25.

Larson, A.M. and F. Soto. 2008. 'Decentralisation of Natural Resource Governance Regimes', *Annual Review of Environment and Resources*, 33(1): 213–239.

Lenaerts, K. 1993. 'The Principle of Subsidiarity and the Environment in the European Union: Keeping the Balance of Federalism', *Fordham International Law Journal*, 17(4): 846–895.

Ministry of Environment, Forest & Climate Change, GoI. 2006. *National Environment Policy*. New Delhi: GoI.

Ministry of Environment Forests & Climate Change, GoI. 2015. *State Action Plan on Climate Change*. New Delhi: GoI. http://envfor.nic.in/ccd-SAPCC (accessed on 02/06/2018).

Ostrom, E. 2010.'Beyond Markets and States: Polycentric Governance of Complex Economic Systems', *American Economic Review*, 100(3): 641–672.

Press Information Bureau, GoI. 2009. PM's address at the National Conference of Ministers of Environment & Forests. New Delhi: GoI; Aug 18. http://pib.nic.in/newsite/erelease.aspx?relid=51926 (accessed on 02/06/2018).

Prud'homme, Remy. 1995. 'The Dangers of Decentralisation', *The World Bank Research Observer*, 10(2): 201–220.

Rabe, B.G. et al. 2006. 'State Completion as a Source Driving Climate Change Mitigation', *NYU Environmental Law Journal*, 14: 1–53.

Revesz, R. 1996. 'Federalism and Interstate Environmental Externalities', *University of Pennsylvania Law Review*, 144(6): 2341–2416.

Shobe, W. and D. Burtraw. 2012. 'Rethinking Environmental Federalism in a Warming World', *Climate Change Economics*, 3(4): 1250018.

Singh, Avantika. 2017. 'Gendering Climate Change Vulnerability: Case Study of Two Villages in Uttar Pradesh', Unpublished Ph.D. Dissertation, University of Delhi.

Srivastava, Nidhi and P.G. Dhar Chakrabarti. 2015. *Green Federalism: Experiences and Practices*. New Delhi: TERI Press.

WMO. 2017. *WMO Statement on the State of the Global Climate in 2017*. https://public.wmo.int/en/media/press-release/2017-set-be-top-three-hottest-years-record-breakingextreme-weather (accessed on 02/06/2018).

Part III

CHANGING DYNAMICS OF FISCAL FEDERALISM

8

REGIONAL BUSINESS CHAMBERS AND FEDERALIZATION IN INDIA

Himanshu Roy

This chapter presents an analysis of the role of regional business, collectively represented by the trade associations and different regional chambers of commerce and industry, in the federalization of India, since 1887, through the creation of linguistic provinces in the pre- and post-independent India and through the fiscal and administrative reforms and political accommodation. The chapter also indicates an emerging subterranean trend of gradual marginalization of regional parties that has played a stellar role in the federalization process in postcolonial India. The new social milieu of globalization that propels standardized attributes of market in different regions drove the regions closer to each other with similar traits, making the regional parties with their unique identities irrelevant. Simultaneously, the expansion of regional business in the pan-Indian market and ventures with global capital impacted its relations with the regional parties. The gradual expansion of the Bharatiya Janata Party (BJP) substituting the regional parties reflects this emerging trend.

Role of chambers: colonial backdrop

The chamber of commerce in India was first established in 1834 in Calcutta, after the British East India Company's external trading right was liquidated by the British parliament in 1833. It was followed by the formation of the Madras and Bombay chambers of commerce in 1835 and 1836, respectively. These chambers were of European merchants formed to protect their trading interests under the new circumstances of free trade. The first Indian merchants' chamber, different from theirs, was formed in 1883 in Coconada. It was called the Native Merchants Chamber.[1] It was followed by the formation of Bengal National Chamber of Commerce and Industry (BNCCI) in 1887. Its constitution was drafted by A.O. Hume and was revised by W.C.

Bannerjee, the founders of the Indian National Congress. Subsequently, in the decades to come many other chambers were formed; most prominent among them were the South Indian Chambers of Commerce and the Andhra Chamber of Commerce.

Actually, the regional chambers, in need of protective market and other requisite support for their business from their local governments, believed that linguistic provinces would provide them the required support in their business expansion, because most of the chambers, in the initial years of their formation, had emerged out of the dominant linguistic regions of the time. It was reflected in their public representations and in mass support to the movements that demanded the creation of linguistic provinces or sought the annulment of partition, as in the case of Bengal in the early decades of 20th century. For the objective of the regional chambers was

> to aid and stimulate the development of commercial enterprise in the regions and to protect the commercial interests of all persons trading there in. . . . For example, the BNCCI had intended from the very outset that . . . the chamber should all along regard itself as virtually interested in the economic development of Bengal in a manner which would not be possible through unqualified membership, as businessmen with their principal centers outside Bengal could not devote concentrated attention and enthusiasm for the development and economic welfare of this Province. It is this limitation of its scope which imparts a distinctive strength to the chamber from the point of view of this state.[2]

Following this objective, it uncompromisingly opposed the partition of Bengal in 1905. In fact, at the planning stage itself, when the partition was being proposed in 1904, the chamber had opposed the idea. Subsequently, it supported the anti-partition agitation, which was guided predominantly by its economic consideration (ibid.: 113). The agitation and the role of the chamber in it facilitated two things later on: (a) a beginning of annual sessions of the Indian Industrial Conferences, which were organized along with the annual sessions of the Indian National Congress since 1905 and (b) a change in the industrial policy of the Government of India, which began to offer "help and guidance" through provincial governments to native business in areas such as handloom weaving and leather processing. The Madras government in particular provided considerable impetus to the application of this new policy. These developments prompted the formation of the Southern Indian Chambers of Commerce in 1909 and subsequently in the formation of other such chambers in the later decades, which played a catalytic role in the federalization process premised on the creation of linguistic provinces. These chambers were consulted by the governments to seek their views on business policies that were also granted a constitutional right to send their representatives to the legislative councils to represent the collective interests of the business, which, in turn, obviously sought conducive business policies from the government.

Simultaneously, the chambers applied collective pressure on the Congress, particularly after 1905, to influence it to accept the principle of federalization (in 1916) and of linguistic provinces (in 1920) as the basis of reorganization of administrative units of the country and of constitutional divisions of powers between the federating units as the mode of governance. Jolted by the 1909 Act, which had fueled separation of its social constituents, the Congress was afraid to part with any further social blocks. The prevalent latent current in the legislative councils also prompted it to accept this new arrangement. This can be judged from the fact that immediately after the consent of the Congress, the Madras, Bihar and Orissa legislative councils passed the resolutions in 1921 exhorting the governments to reorganize the administrative units on the basis of linguistic provinces. In 1927, when the Simon Commission arrived in India, the government of India and the India office were flooded with the memorandum to be forwarded to the Simon Commission for "the amalgamation of the Oriya speaking people" and for "the formation of separate Andhra, Tamil and Kannada provinces." Subsequently, resolutions were "moved in the Central legislature for the formation of separate Andhra, Tamil and Kannada provinces, and a resolution urging the formation of a separate Andhra province was adopted in the Madras Legislative Council" (Ghosh 1996: 17).

The chambers' role in the federalization process further accelerated in the 1940s. The Southern Indian Chamber of Commerce strongly recommended to the Industrial Planning Committee of the Madras government that economic planning should be primarily provincial in nature and the government should "do everything to improve the trade and industries and help to retain as big a slice of them as possible for the benefit of the sons of the soil" (ibid.: 20). The chamber was dismayed to observe that the citizens of Madras province were being outwitted by nationals of other provinces in the matter of the race for industries and enterprises and demanded in a resolution of industrial and commercial conference that "regional interests should be safeguarded" (ibid.: 20–21.) The Bengal National chamber, too, had put forth its own regional plan for the development of its province. The countervailing force to the regional interests was the presence of the Federation of Indian Chambers of Commerce and Industries (FICCI) representing the collective interests of big business, which itself had emerged out of the regional chambers of commerce in 1927 and was now against their dominant roles in the regional market. It opposed the principle of regionalization of industries and urged that private enterprise should be given the full freedom and facilities to establish industries in places most suitable for such development. It regretted that every time a proposal to establish an industry in the Madras presidency was given, it was turned down on the ground that the firm sponsoring the establishment of that industry did not belong to that presidency. It complained that the present policy of regional development would lead to the creation of monopolies for persons or firms born only in a particular district of province.[3] The FICCI was afraid that under the Cabinet Mission Plan, which was premised on the classical concept of federalism envisaging larger powers for federating units and lesser powers for the center, the provinces "will exercise a larger measure of autonomy on all matters pertaining to the province. All most all the aspects of trade and commerce will thus be exclusively provincial spheres."[4] It urged the central government

> to include special provisions in the Constitution Act . . . to provide that any unit, province or State shall not have power to pass any laws or take any executive action prohibiting or restricting the entry into or export from one unit or province to another of any goods of any class of description. It should similarly be laid down that no unit or province should be entitled to impose any tax, cess, toll or due which, as between goods manufactured or produced in one province or unit and similar goods not so manufactured or produced, discriminate in favour of the former.[5]

This demand was urged despite a major shift in the central government's economic policy in 1945 that had transferred the control of 20 major industries from the purview of provincial governments to the central government.[6] More, when the Constituent Assembly was formed, the big business lobbied through different constitutional committees and subcommittees to put more industries/business sectors under the purview of the central government. Yet despite such acts, it was so afraid of the power of the regional businesses that it did not let pass any opportunity that struck at their interests. And in its role, it was ably supported by the political leadership that was best expressed by the Nehrus themselves. In 1928, for example, while presenting the Swaraj constitution for the country, Motilal had envisaged a unitary form of government that had nullified all the struggles of regional businesses till then for the federalization process that the Congress in principle had agreed to. Later, in 1948, addressing the Annual meeting of the FICCI, Jawaharlal had observed that "more dangerous than communalism is provincialism. And if every province formulates its own economic policy and adds new obstacles then it is unwanted."[7] The interrelations of business and politics in the quest for a monopolistic market and absolute power reached its crescendo in the last months of 1946 and early months of 1947 (Menon 1957: 358–359) when an unrelenting Jinnah and Muslim chamber of commerce, in the quest for their own power and market, not conceded by the Congress and big business, actuated partition of the country. The Congress propelled by the FICCI used this situation for their respective benefit. They agreed to partition the country, and then, on the pretext of curbing the fissiparous trend, appropriated all the major political–economic powers of the provinces and put them under the purview of the central government. The structure of powers of the federating units as envisaged by the Cabinet Mission Plan was radically altered forever under the new constitutional scheme after partition. The regional businesses' dominant role in the provinces, the kind of which existed in the pre-partition years, ceased to exist in independent India. Their system of representation in the legislatures, the constitutional right earned after a tenuous struggle against the colonial state, was dismantled forever.

Postindependence

Fortunately, the regional businesses did not give up the fight. They propelled the masses to force the central government for the reorganization of the states on a linguistic basis despite strong opposition of the Congress leadership and of big business, represented by the Linguistic Provinces Commission of the Constituent Assembly, the Linguistic Provinces Committee of the Congress Party and the Citizens' Committee of the J.R.D. Tata, H. P. Mody and Purshotamdas Thakurdas, respectively (Ghosh 1996: 12, 14, 24). For, it may be noted here, big business was never in favor of reorganization of states premised on the linguistic basis after 1927; neither did the political leadership voluntarily weld itself to the principle of federalism and linguistic provinces. It agreed to the cause only under the duress of regional businesses and mass pressure that compelled the passage of different resolutions in legislatures and at the party's forum in pre-partition years and of the constitution of the State Reorganization Committee, finally, in 1953.

The formation of the linguistic provinces, however, did not alter the power structure of 1950 Constitution. Rather, it only created new states which were in tune with the Constitution, with elementary powers to be of little use for regional businesses. It did not alter the dominant position of big business and of the central government acquired after the partition, which had appropriated the residuary and other major powers of the state as envisaged by the Cabinet Mission Plan. Regional businesses in retaliation began to invest in regional parties, which had begun to acquire political power since 1967 after the formation of the linguistic provinces commenced from 1956–57. In the next 30 years, by 1987, major linguistic regions of the country were reorganized into provinces, and their governmental powers gradually came under the control of the regional parties, or the regional parties began to play a major role in the provinces. And in a few exceptional cases, the national parties began to play a limited role of regional parties. In other words, it may be derived that where regional business was dynamic, the role of the regional parties had been more assertive in the provinces (Baru 2001: 226), which is best reflected, four decades later, in the dismantling of the one party-dominant system and of the emergence of a bi-coalitional system at the federal level in which the regional parties play an assertive role, and in the emergence of the bi-party, bi-coalitional system existing at the provincial level (Roy 2004). In brief, the Congress, which was committed to the unitary system as an ideal model of polity and accepted the principle of federalism premised on a linguistic basis only under duress, was itself thrown from the pedestal by the regional forces in the course of development. The dominance of the Congress party was dismantled forever, and it was substituted by an alternative national/regional party systems at the center and in the provinces. More, in a twist of history, regional businesses, which were seeking a place under the sun from the Congress, initially themselves became a part of global business under the growing liberalization process of capital unleashed by this party. They transcended their regional character when opportunity beckoned and transformed themselves into a pan-Indian, pan-global group. Their entry into the top 50 business houses reflects their transformation and is proved by many case studies. Sanjay Baru (2001: 216–225), for example, brings to notice how small business organizations (Nagarjuna, Dr. Reddy's Lab., etc.), with the help of their local governments, local financial institutions and in collaboration with transnational corporations, developed themselves to be in league with big business. Their struggle for and success in the termination of the license-permit raj, which was more favorable to big businesses, facilitated such development, and in the process, their organizational structure, expansion and functioning and their market share and visibility underwent transformation, and their development, simultaneously, propelled the establishment of regional stock exchanges and ensured their vibrancy. The regional business firms became a part of the listed companies on these stock exchanges and then catapulted themselves to the national stock exchange.

But as the market economy expanded and deepened, new subregional chambers emerged within the existing linguistic provinces demanding new states or autonomous councils contesting the old regional chambers, which had now acquired a pan-Indian market and presence in the global market.

Emerging trend

Politically, all these developments had their impact on regional politics, which varied in different regions in different degrees. As the old regional chambers, after the formation of the linguistic provinces, became more engaged with their issues of fiscal autonomy, routine governance, tax relief and market expansion, the role of the regional parties changed. It became more engaged with electoral politics and with the formation of the government. Its raison d'être as an autonomous regional entity, therefore, ceased to exist. Its survival against the Congress party as the leader of the regional movement and of the regional identity was no longer tenable. Its role of governance, therefore, was gradually taken over by the BJP, which has emerged, in the meanwhile, as the new pan-Indian party.

Under the new social milieu propelled by globalization, a new trend is witnessed that indicates the gradual substitution/marginalization/co-optation of regional parties in different states that had provided alternative to the Congress system after 1967. The new emerging alternative to the Congress party at the pan-Indian level is the Bhartiya Janata Party, which has intruded into government formation or has begun to play an important role in it in the provinces from Goa to Arunachal and from Karnataka to Jammu and Kashmir. Although it has yet to prove itself as a pan-Indian party on a long-term basis, nonetheless, it has become a new alternative political mascot of global capital in India that propels standardized/unformal attributes in different regions while co-opting in itself the sanitized versions of the regional specifics.

In the economy, simultaneously, the capitalist development reflects that, in the course of expansion of the market economy, a section of regional businesses has grown to become part of big business, and a large section gets the opportunity of accessibility to capital, market and technology to become a partner in global business. Their elephantine expansion in a short duration reflects their success and also gets reflected in the composition of the Confederation of Indian Industries (CII) and of the Associated Chambers of Commerce (Assocham) (Baru 2001: 225, 230), which are different from the old big business of the FICCI. In fact, thanks to the increasing expansion of this economy and availability of opportunity for vertical mobility to different sections of society, there has even emerged the formation of Dalit Chambers of Commerce and Industry (DCCI). More recently, the regional business chambers,[8] despite hiccups, played important roles in the creation of the Goods and Services Tax (GST). Without their support, it could have been difficult for the federal government to actuate one unified tax regime for the entire country.

Twenty-five years after the beginning of globalization, the global and big Indian business propelled the government to initiate this process. The existing tax regime, cumbersome and different in every province, had irritated businesses of late. Moreover, the standard operative procedures of the businesses of big companies across India, facilitated by technology, had already laid the groundwork for a uniform tax regime. Therefore, it became easier for the federal government to initiate and complete the process within such short duration.

In the backward regions, particularly in the hilly and tribal regions, the standard global attributes of market are yet to deepen; it encourages local subregional chambers and local tiny parties/political groups, nongovernmental organizations that cater to the local requirements.

Conclusion

Apparently, it may seem dichotomous at this political juncture to note the gradual marginalization of the regional parties. But a rigorous analysis of the emerging economic trend and of the party systems in the provinces indicates the subterranean changes underway. The universal standardized attributes of the cosmopolitan production bases of the companies in the economy under globalization have impacted the culture and politics of the day, resulting in the co-optation of varied regional political and economic issues in its governance program by the BJP, the alternative political mascot of the global capital that is expanding its wings in the regions. Its refutation is possible if, alternatively, the regional parties themselves act to represent the global capital in the regions. Both these developments, however, are yet to be proved on long-term basis.

Notes

1 The Native Merchants Chamber was formed by local merchants of the subregion of Andhra Pradesh when they were denied membership to the local European chamber at Kaakinada, which is a seaport, 175 km away from Vishakhapatnam, toward Chennai by road, and is directly connected with international trade.
2 Bengal National Chamber of Commerce and Industry, Souvenir Volume 1887–1962 Calcutta, pp. 39, 49.
3 Ibid.
4 Federation of Indian Chambers of Commerce, New Delhi, Proceeding of the 20th Annual Meeting, 1947, Volume 3, p. 86.
5 Ibid.
6 Ibid., Volume 1, 1946, p. 128.
7 Ibid., Volume 3, p. 29
8 Recently, in October 2016, I visited the office of Native Merchants Chambers at Kakinada, which is the earliest Indian business chamber. It is surprising to note that a "native" chamber of local merchants was formed at such a nondescript trading center as a protest. These merchants were engaged in the export of coconut filaments and other related products and used to seek tax concessions and other trade benefits from the colonial administration.

References

Baru, Sanjay. 'Economic Policy and Capitalism in India', in *Transforming India*, eds. F. Frankel et al. Oxford: Oxford University Press, 2001, p. 226.

Ghosh, Suniti Kumar. *India's Nationality Problem and Ruling Classes*. Calcutta, 1996, p. 17.

Menon, V.P. *Transfer of Power in India*. Calcutta, 1957, pp. 358–359.

Roy, Himanshu. 'Party Systems and Coalition Politics in Indian States', in *Coalition Politics in India*, eds. M.P. Singh and Anil Mishra. New Delhi: Manohar, 2004.

9

GLOBALIZATION, THE WELFARE STATE AND COOPERATIVE FEDERALISM IN INDIA

Alisha Dhingra

Historically, the concept and practice of cooperative federalism and welfare state emerged hand in hand. The concept of "cooperative federalism" first appeared in the terminology of comparative federal praxis as a response to the doctrine of dual sovereignty that was prevalent in the constitutions of the United States, Canada and Australia. The social security program that was initiated in the United States by President Franklin D. Roosevelt under the New Deal package was followed by more extensive welfare policies in Canada and Australia. The success of these policies necessitated strong cooperation among the federal government and provincial governments. The transformation from a laissez-faire noninterventionist minimal state to a welfare state was accompanied by the transformation from the doctrine of dual sovereignty to the praxis of cooperative federalism. A. H. Birch defines cooperative federalism as administrative cooperation between the central government and state governments and the partial financial dependence of regional governments upon federal government (Birch 1955: 305)

When the Indian Constitution was negotiated, the framers had before them the model of federalism in the United States, Canada and Australia. The framers of the Indian constitution went beyond the constitutions of these nations and also learned from their praxis. Thus, even while adopting federalism, they avoided the dual sovereignty between the center and the states. One of the prime tasks of the postcolonial state in India was to lift the masses from poverty and destitution. This surely required a welfare state and cooperation between the center and the states in the pursuit of development. Thus, in the Indian Constitution, cooperative governance/federalism is apparent in both the political and the fiscal sphere. Austin emphasizes that the Constituent Assembly of India was possibly the first constituent assembly to hold onto the conception of "cooperative federalism" from the beginning itself (Austin 1972: 232). The model adopted by the constitution makers in India was very different from Wheare's definition of federalism: "The general and regional governments of a country shall be interdependent of each other within its sphere" (Wheare 1956: 97). This definition of Wheare was derived from the federal systems of United States or Australia, and therefore, Wheare described Indian governance as "quasi federal." However, Austin believes that Indian federalism was a kind of cooperative federalism that has produced a strong central government but that does not lead to conversion of state governments as administrative agencies for policies of the central government (Austin 1972: 232). Rather, what we see in India is the increasing interdependence of central and state governments that underlines the spirit of cooperative federalism.

Today, there is a renewed emphasis on cooperative federalism by the current prime minister. The new economic reforms that were to transform the welfare state into a neoliberal state had many hostile economic and social costs. Thus, we observe that though there was a cut in certain welfare programs, it went hand in hand with the expansion of certain other welfare programs and even the introduction of new welfare policies.

From a comparative perspective, India's overall social spending remains low. For example, India's expenditure on health is only 4 percent of GDP, compared to 8.8 in South Africa and 9.3 in Brazil. The expenditure on education is also dismally low, standing at 3.4 percent of GDP, while South Africa spent 6.6 percent and Brazil 5.8 percent. Though low from a comparative perspective, India's total social sector expenditure has been on the rise. What is imperative to note is the fact that the central government's share of the total expenditure has declined in recent years. The overall boost in social expenditure is driven by the rising proportion of state-level expenditure on the social sector. This justifies the significance of the states as the vital arena for understanding welfare policies in India (Tillin et al. 2015: 11).

Impact of globalization on Indian federalism

Globalization has expanded the conception of federalism to include both vertical federalism and horizontal federalism. Vertically, the chain involves central government, state governments and local governments. Moreover, if we scale one level up in the chain of vertical federalism, we can also include various international agencies. The international organizations such as the World Trade Organization (WTO) and World Bank have been constantly impacting center–state relations in India. We can also see the involvement of international NGO's in the implementation of Centrally Sponsored Schemes. Horizontally, the chain of federalism includes business associations, NGOs and local communities. Moreover, there exist linkages between the chain of vertical federalism and horizontal federalism. For instance, NGOs cooperate with both international agencies and state governments. Another instance is the involvement of business associations in interactions between the central government and international organizations such as the WTO. Business houses and associations have become part of the official delegation in the ministerial meetings. Traditionally, federalism was considered as division of powers in the policy domain between the union government and state governments. What we observe in the age of globalization is the inclusion of NGOs (national and international), international organizations (United Nations, WTO, World Bank, etc.) and business associations (Confederation of Indian Industry [CII], Federation of Indian Chambers of Commerce and Industry [FICCI], Associated Chambers of Commerce and Industry (ASSOCHAM)) in the policy domain.

The onset of globalization in India became noticeable when India boarded the path of liberalization in July 1991. Linda Weiss argues that rather than viewing globalization as a technologically induced phenomenon, we need to emphasize its political rationale. This rationale can be understood from the fact that globalization is ultimately an outcome of governments, either readily or reluctantly, yielding to the pressure from financial interests. Moreover, in many cases, governments are in quest of facilitating rather than constraining the internationalization of corporate movement (Weiss 1997). The central government introduced economic policy reforms leading to a silent economic revolution or, as Rob Jenkins puts it, "reforms by stealth" (Jenkins 1999). Jenkins point out that federalism eased the political sustainability of India's reform process by relocating opposition to the state level (ibid). It needs to be emphasized that although structural adjustment reforms were introduced by the central government, their implementation necessitated the efforts of the state governments. It is interesting to observe that significant sectors such as the industrial infrastructure, power, agriculture, education, etc., fall under the jurisdiction of states, and the new economic reforms required significant changes in these sectors. Moreover, with the dismantling of centralized regulations and controls, there has been significant devolution of power to the states. Thus, it becomes necessary to gain the support of state governments, which becomes decisive for the successful execution of reforms (Pai 2013: 12).

Various scholars have tried to comprehend the far-reaching character of changes in India's policy regime in the context of center–state relations. Lawrence Saez has observed that economic liberalization policies initiated in the 1990s impelled modification in federal relations from intergovernmental cooperation toward interjurisdictional competition among the states (Saez 2002: 135). N. Echeverrigent has observed considerable "decentering in the economic relations between regional states and the centre" (Echeverri-Gent 2002). Moreover, it has been argued that the federalization and neo-economic reforms are mutually reinforcing and the autonomy of both the state governments and private sector has enlarged (Dua and Singh 2003: 186). In the newfangled political context, express governmental control of the economy has been replaced by autonomous regulatory bodies, giving rise to a fresh trend of a regulatory state and "sectoral federalism," which is more horizontal than vertical owing to the nature of Indian economy (ibid). We can also trace interlinkages between another important dimension of federalization, that is, decentralization and new economic reforms. Robert Jenkins has emphasized the autonomous political logic of the structural adjustment program initiated by the Narsimha government. He argues that both the populist policies introduced by the Indira Gandhi government and new economic reforms and decentralization introduced by the Narsimha government facilitated the recovery of the lost strength and vivacity of the Congress. Decentralization helped in easing the pressure for the central government and pushed down the opposition to the reforms at the levels below the central government (Jenkins 1995).

India has become significantly more federalized than before since the 1990s. This is manifest from the enlarged political autonomy of states and the increase in revenue of state governments vis-a-vis the union government; the mounting maneuverability of the constitutional heads of state – president and governors; the rising salience of union constitutional agencies like the Election Commission of India and intergovernmental agencies like the ISC (Inter-State Council) and NDC (National Development Council); the reduction in the role of centralized planning; the formation of a string of independent regulatory agencies and appellate tribunals under parliamentary statues in crucial sectors of the economy such as communications, electricity, insurance, finance, the stock exchange, etc.; and the constitutional entrenchment of local self-governing institutions of *panchayats* and municipal corporations, etc. (Singh and Saxena 2003).

Since the 1990s, scholarly attention has shifted from an examination of the continuing process of reform at the central level since 1991 to exploring the interface between economic reform and democratic politics in the states (Pai 2013). Various scholars examined the diverse manner in which particular states shape and execute economic reforms. There has been an emerging argument that the policies initiated by the central government are not the only determining factor for the success/failure of development strategies in India. Due to the existence of the federal structure, political choices made by politicians at the state level and the regional political competition are crucial variables in determining the success/failure of development strategies (Jenkins 1999; Sinha 2005).

Rudolph and Rudolph have perceptively noted that economic reforms in India since 1990s have transformed India from a "command economy to federal market economy" facilitating a wider "sharing of sovereignty" within the state and the market. In this framework, the states hold a greater segment of economic sovereignty than they held during the centrally planned economy. The economic performance of the states now depends more than ever before on their own actions. The states have turned out to be the primary field for private investment, for which they need to improve the policy milieu, ensure good governance and develop infrastructure and human resources. Moreover, the states are confronted with crucial challenges posed by liberalization, which they need to address to ensure the success of economic reform such as deficits, user charges and administrative pricing (Rudolph and Rudolph 2001).

Several scholars have argued that states in the Indian federation have become assertive since the late 1980s and have created a new space for themselves. The idea of states as independent loci of development is reflected in the arguments of globalization theorists like Kenichi Ohmae, who believed that states/subunits of nation in the era of globalization are to be seen as budding engines of enormous regional economic growth (Ohmae 1995). In the Indian context, policy makers hold that the process of liberalization juxtaposed with the rise of coalition forces, a powerful judiciary and the social movements, etc., have tilted the balance toward a more decentralized federalism in which the states have a renewed concern for development within its spheres (Bhattacharyya 2009). This development been accredited by scholars to the intermeshing of several factors, such as the onset of the coalition government, the emergence of a multi-party system at the union level and economic decentralization resulting from the process of globalization (Saez 2002; Singh 2003; Bhattacharyya 2009; Saxena 2016).

Different state governments have recognized the benefits of the new economic reforms with the onset of globalization. Subrata Mitra has argued that as the central government has sacked its convoluted regulations, the role of the states in the Indian federation has been reoriented (Mitra 2007: 86). In this context, Rudolph and Rudolph have analyzed the transformation of the Indian state "from an interventionist state to a regulatory state" (Rudolph and Rudolph 2001b). Saxena (2016) argues that consequent to the new economy reforms, economy has opened up and became more competitive. This has reduced the role of the central government as the only authority of regulation, leading to empowerment of the states to chart out their own path and pace of development. Moreover, she argues that this newfound relative economic independence has imbued a sense of confidence among the states in the domains of society, culture and polity. This has also brought to the forefront the study of state politics in India. The politics of Indian states and identities within their sphere have emerged as an independent subject of analysis (Saxena 2016).

Saxena (2016) also points to a line of caution that the newfangled freedom of action provided in the era of globalization has advanced only a few states in India, and the rest of the states are unlikely to obtain advantages because of their failure to attract investment from private national and international market consequent to a host of factors such as social and cultural heterogeneity, the small size of the states and the unfavorable topographical location. Another major challenge in the era of globalization is protection of the micro-minorities, discrepant majorities and minorities within minorities (internal minorities within all castes, tribes, languages and religions). It is difficult for the state to provide special benefits to these groups because they are scattered all over the territorial scope of the Indian federation. This challenge has become convoluted considering the concurrent effect of globalization, regionalization and localization of identities in India (ibid).

One significant influence of globalization on Indian federalism has been a manifest intensification in regional disparities. Francine Frankel has highlighted the propensity of India moving in the path of a "dualistic economy" since the early 1990s because the implementation of neoliberal economic reforms has been done without redistributive change. Frankel exemplifies the intensification of interstate disparities, with Maharashtra and Gujarat demonstrating the prospect of reproducing the East Asian levels of annual growth in the 1990s. The states of Madhya Pradesh, Tamil Nadu and West Bengal are rising above the national average, whereas states like Bihar, Uttar Pradesh, Orissa and some richer states of Haryana and Punjab have experienced a deterioration in annual growth rates (Frankel 2005: 604–605).

B. B. Bhattacharya and S. Sakthivel opined that in the reform years, on the one hand, advanced and highly industrialized states have observed a big jump, and on the other hand, backward states having higher population growth are not capable of attracting investment (both public and private) because of a number of factors such as low income and poor infrastructure coupled with poor governance (2007: 475). Amaresh Bagchi and John Kurian have noted that one of the immediate bases of the broadening regional disparities during the '90s was the patchy movement of investment to several states after neoliberal reforms (Bagchi and Kurian 2005: 336). They have analyzed the percentage share of investment tenders obtained by different states between 1991 and 2000. The percentage for Group I states comprising Andhra Pradesh, Haryana, Gujarat, Karnataka, Maharashtra, Kerala, Punjab and Tamil Nadu was 66.7 percent, whereas the percentage share of Group II states, covering Assam, Bihar, Orissa, Madhya Pradesh, Rajasthan, West Bengal and Uttar Pradesh, was 27.4 percent. In terms of foreign direct investment (FDI), five states – Gujarat, Andhra Pradesh, Karnataka, Tamil Nadu and Maharashtra –accounted for approximately 75 percent of the total foreign direct investment (FDI) obtained in India since neoliberal reforms (Bagchi and Kurian 2005: 336–339).

Similarly, Bhattacharyya and L. Konig argue that globalization and neoliberal reforms have ushered in extreme inequality among states with regard to human development. There is huge disparity in the amount of FDI received by "forward states" and "backward states" (Bhattacharyya and Konig 2016: 145).The creation of Special Economic Zones (SEZs) in the 2000s in states is one of the most controversial policies initiated in the era of liberalization. Jenkins analyzes the response of various states to the oppositional movements that emerged as a reaction to SEZs. He argues that during the period of liberalization, we witnessed widespread opposition to unfair laws and practices in the states leading to social legislation such as the Mahatma Gandhi National Rural Employment Guarantee Act (MGNREGA), the Right to Information Act (RTI), the National Food Security Act (NFSA) and the Right to Education Act (RTE) (Jenkins 2013).

On the other hand, Nayar points out that globalization alone cannot be held responsible for creating inequalities. He believes that the policies initiated by state governments and the working of institutions are responsible for differential development across states. Nayar views globalization as a positive force that has diluted central controls, and the initiation of tax reforms have built up a national market and engendered revenue for welfare programs such as MGNREGA, fashioning political stability and forging national integration. He points out that in some states such as Uttar Pradesh and Andhra Pradesh, economic reforms have created dissatisfaction leading to demands for the creation of small states, which would help in creating a more composed, multipolar federation (Nayar 2013).

Bhattacharyya and Konig (2016) believe that Indian federalism is transforming from "cooperative federalism" to "competitive federalism". Sinha (2004) argues that since the 1990s, the character of competition changed from "vertical" (in which states vie with each other for resources from the center) to "horizontal" (in which states vie with each other for resources from varied actors). What needs to be noted is that although the number of contending states has multiplied and there is a seeming conjunction of policy and discourses, this has not led to the reduction of regional disparities on outcomes and institutions; rather, these disparities have increased since 1991 (ibid).

Another interesting trend is that the "re-regulation of liberalization by many state-level officials" has boosted the role of states in investment policies. Several states regarded neoliberal reforms as a prospect to reaffirm their own regional agenda and extort more regional autonomy in development policy. States such as West Bengal, Tamil Nadu, Andhra Pradesh and Madhya Pradesh tried to reregulate economic reforms for their specific interests (ibid).

In the pre-liberalization era, competition among states was "indirect, vertical and asymmetric." Owing to "the regulatory licensing regime that allocated both public and private goods, there was competition between the states for these allocations, but the competition was mediated through the central government" (Bhardwaj 2015: 116). This is not to refute that even in pre-liberalization era, a few states such as Gujarat tried to lessen dependency on central government by pleading with the Guajarati Diaspora settled in different countries to invest in Gujarat. A few other states such as Maharashtra attempted to follow the same line. But what needs to be emphasized is the fact that such measures and competition were limited to only a small number of states. Post-1991, industrialized states such as Gujarat and Maharashtra have to labor very hard in order to hold on to their first and second positions as investment destinations, as the competition became "wider and symmetric," with states such as West Bengal, Andhra Pradesh, Orissa, Karnataka, Delhi, Goa, Bihar, Uttar Pradesh and Mizoram making efforts to attract investment. State leaders are also dealing with multilateral organizations such as the World Bank for loans (ibid: 120).

Welfare policies and neo-liberal reforms in states

Neoliberal reforms, contrary to popular perceptions, do not essentially shrink the range of political action but could often be employed to both inflate political authority and fortify support bases. Political agents can often maneuver markets in public services to generate varied outcomes. Policy reforms in different public services can differ in terms of costs levied (Gingrich 2011). Richard Snyder's research on the deregulation of coffee in Mexico found that neoliberal reforms, while unleashing market forces, could also elicit what he calls "reregulation," where political incumbents employ the new prospect to both conserve power as also to enlarge it (Snyder 2001). Gupta and Sharma have argued that postcolonial contexts may provide us rationale to review generally received descriptions of neoliberal states. In a postcolonial country, where high rates of poverty coexist with high rates of growth (in a neoliberal economy), what we observe is not the ending of welfare programs and its alteration with workfare programs but the concurrent extension of both kinds of programs. They argue that though market reforms may possibly have an enormous impact on some bureaus of the state at the federal level, their sway on lower levels of government and on agencies not straightforwardly associated with industry or consumer goods is much less apparent. They call for a disaggregated approach to the state that looks at it in a frame which makes it easy to perceive that key policy shifts at the federal level do not essentially transform the structure at the lower levels of the state. They emphasize studying everyday dealings of specific branches of the state to comprehend what has in reality altered, and at which levels, and to seek explanation for the circumstances in which discrepant representations of "the state" circulate. (Gupta and Sharma 2006).

We need to bridge the divide between the study of public policy on the one hand and the study of political parties, electoral politics and federalism on the other hand (Tillin et al. 2015: 6). There has been an argument made by scholars around the world about the increasing ideological congruence between political parties reducing the significance of elections in influencing public policies (Boix 1997; Castles and Mckinlay 1977, Yadav and Palshikar cited in Tillin et al. 2015: 6). More recent research on explaining levels of expenditure on public goods in different states across India has suggested that factors associated with the election cycle or the effectual number of political parties contesting for power in a given state is more significant than partisan or ideological factors (Chibber and Nooruddin 2004; Saez and Sinha 2010, cited in Tillin et al. 2015: 7).

The formulation and implementation of welfare policies across Indian states depends upon a range of variables such as long term paths of historical development and the ability of the leader to initiate breaks; the pattern of electoral competition; the nature of administrative structures and agency internal politics; the association between state and non-state actors in policy making, policy execution and policy analysis; and the embeddedness of state governments in a string of exterior relationships within a multilevel governance structure that plays a vital role if not a deterministic one in shaping their policy and fiscal place. This string might constitute the central government, other state governments and international organizations, as well as other foreign governments. Thus, we cannot view the states as an independent unit of analysis but understand the interdependence of the state on other levels of the federal and international system (Tillin et al. 2015: 23–26).

In a multilevel federal system, we need to study "the relationship between political environment and social policies across Indian states" (Tillin et al. 2015: 2). This would help us to decipher variations in public policy formulation, implementation and monitoring across states. Significant variation exists between states in terms of "historical patterns of socio-political dominance, the organization of political parties and interest groups, bureaucratic structures, economic context and the agency of political leaders" (ibid.). The experiences at the state level reveal that the relationship between growth and inclusiveness is not essentially a trade-off. There is considerable diversity in how state governments execute, influence and add to policy agendas as they try to maintain equilibrium between their approaches to economic development and social policies while maintaining democratic legitimacy (ibid.). Tillin et al. (2015) argue that the overall social policy mix varies across states, resulting in different types of emergent "welfare regimes" as states tackle with analogous tribulations in diverse ways; undertake different decisions about the execution of centrally designed programs; and have different needs resulting from the prevailing patterns of socioeconomic development (ibid.: 26).

Rob Jenkins argues that politics – that is, "the interplay of actors, institutions, and ideas in the pursuit of power" – is a constant influence throughout the "process of designing, establishing, and operating any government programme" (Jenkins and Manor 2015: 168). Politics influences how resources are allocated, suspended or stolen; the degree to which rules and norms are followed; and who benefits, who loses and who ultimately governs. This path of causality is not a one-way lane. The success of welfare programs provides poor people with material resources, which allow them to augment their rights, capabilities and influence. Successful programs lead to augmentation of a government's legitimacy and popularity, and failure of these programs often undermines them. Jenkins argue that the effects of any social programs can modify formal institutional relationships and the informal rules of the game for years to come. In their study of MGNREGA, Jenkins and Manor emphasize the role of "political agency" --the eagerness and ability of elected leaders to shatter path dependency. Political actors can willfully choose to discard past legacies or to reframe them creatively for new objectives. Moreover, they assert that politics and political agency control MGNREGA's operation at all levels of the political system, from central government to the state capitals to the village *panchayat*. They regard the emergence of the village *panchayat* as an effective political institution as one of the most perceptible legacies of the functioning of MGNREGA. They highlight the impact of numerous state level factors on everything from the blueprint of administrative rules and structures for executing MGNREGA to the relationship between government officials and civil society organizations (ibid.:169).

Kailash and Rasaratnam (2015) argue that the states need to be considered as crucial locations of policy making as much as they are seen as arenas of electoral competition, caste mobilization and party systems. The modifications of health insurance policy in Tamil Nadu and Kerala substantiates the findings from different corners of the globe, which reflects that liberalizing reforms do not necessarily generate outcomes that privilege the market. The study of health insurance schemes reveals how and why policymakers in Tamil Nadu and Kerala purposely distorted the liberalizing impulse of health insurance to make it work to the advantage of state-run hospitals. First, the public sector became both a "privileged supplier as well as financier of healthcare." Second, public sector heath provision has become an object to be improved rather than an object that has failed. Lastly, the "individual becomes a citizen to be encouraged to use public facilities rather than a consumer entitled to make choices in a competitive market." The course of health insurance in Tamil Nadu and Kerala reflects the significance of state-level politics in determining the content and outcome of public policies that have their genesis at national and international levels. Three variables at the state level – electoral mobilization of backward caste groups, the nature of party systems and the interest of political leaders – influence the operation of policies (Kailash and Rasaratnam 2015: 41–42).

Tillin, Saxena and Sisodia (2015) have compared the politics of food subsidies in Chhattisgarh and Madhya Pradesh and mapped out the differences in terms of role of the chief minister in Chhattisgarh and the lack of it in Madhya Pradesh in reforming bureaucracy and establishing direct relationship with the rural poor. The second variable they study is the degree of interagency cooperation within the bureaucracy: a strong coordination in Chhattisgarh and lack of it in Madhya Pradesh. Third, vigorous participation of civil society workers in Chhattisgarh and the lack of it in Madhya Pradesh. They argue that the significance of these factors needs to be understood in the backdrop of the political economy of these two states. Thus, though both states were governed by the same political party, that is, the

BJP, there were significant variations in political economy and other factors resulting in a differential focus on food subsidies.

Conclusion

Globalization has had a profound impact on Indian federalism. Some scholars have argued that India has moved toward greater federalization (Singh and Saxena 2013) since the 1990s, while others have used terms such as quasi-confederacy (Verney 2003) or a federal market economy (Rudolph and Rudolph 2001a). One of the most visible trends of this era is the rising economic power of the states; the states are not only dependent on central resources, they compete among themselves to attract private investment and foreign direct investment. Also, we witness that the chief ministers of states have become more assertive, many times accusing the central government of discrimination as well as negligence of the interests of the state.

Till the decade of 1980s, state politics was a neglected area of study. With the transformation of the party system in the 1990s, it became essential to shift the focus on electoral politics of the states. The recent literature argues for focusing on the state not only as an arena of electoral competition but also as an arena for public policy formulation and execution. Reflecting this spirit, a recent work (Tillin et al. 2015) has employed the concept of "welfare regimes" in states, offering as evidence various state-level factors such as historical patterns of sociopolitical dominance, the organization of political parties and interest groups, civil society actors, bureaucratic structures, economic context and the agency of political leaders. The literature not only emphasizes the "state" as a unit of analysis but also looks at the embeddedness of states in relationships both above (central government and also international agencies) and below (local governments). Moreover, it also emphasizes on horizontal interaction of the states with NGOs and community.

The outbreak of COVID19 has underlined the significance of cooperative federalism to meet the pandemic crises situation. In federal countries, vertical and horizontal cooperation between different levels of government is essential to deal with such crises. Vertical cooperation and mutual learning between the central government, state governments and district administration is essential, and horizontal learning and cooperation among states and initiatives by the private sector and civil society are the need of the hour during crises.

References

Austin, Granville. 1972. *The Indian Constitution: Cornerstone of a Nation*. New Delhi: Oxford University Press.

Bagchi, Amaresh and John Kurian. 2005. 'Regional Inequalities in India: Pre-and Post-Reform Trends and Challenges for Policy', in Jos Mooij (ed.), *The Politics of Economic Reforms in India*. New Delhi: Sage Publications.

Bhardwaj, Ritesh. 2015. 'Impact of Economic Liberalization Policy on Indian Federalism: The Changing Pattern of Intergovernmental Relations', PhD dissertation submitted to Department of Political Science, University of Delhi.

Bhattacharya, B.B. and S. Sakthivel. 2007. 'Regional Growth and Disparity in India: Comparison of Pre- and Post-Reform Decades', in Baldev Raj Nayar (ed.), *Globalization and Politics in India*. New Delhi: Oxford University Press.

Bhattacharyya, Harihar. 2009. 'Globalization and Indian Federalism: Re-Assertions of States' Rights', in Hans Lofgren and Sarangi Prakash (eds.), *The Politics and Culture of Globalization Indian and Australia*, pp. 99–119. New Delhi: Social Science Press.

Bhattacharyya, Harihar and Lion König. 2016. 'Conclusion: India's Second Tryst with Destiny', in Harihar Bhattacharyya and Lion König (ed.), *Globalisation and Governance in India: New Challenges to Society and Institutions*. London and New York: Routledge.

Birch, A.H. 1955. *Federalism, Finance and Social Legislation in Canada, Australia and the United States*. London: Oxford University Press.

Dua, B.D. and M.P. Singh. 2003. *India's Federalism in the New Millennium*. Delhi: Manohar Publications.

Echeverri-Gent, John. 2002 'Politics in India's Decentered Polity', in Alyssa Ayres and Philip Oldenburg (eds.), *India's Briefing: Quickening the Free of Change*, pp. 19–53. New York: M. E. Sharpe.

Frankel, Francine. 2005. *India's Political Economy 1947–2004*. New Delhi: Oxford University Press, 2nd edn. (orig. 1978).

Gingrich, Jane. 2011. *Making Markets in the Welfare State: The Politics of Varying Market Reforms*. Cambridge: Cambridge University Press.

Gupta, Akhil and Aradhana Sharma. 2006. 'Globalization and Postcolonial States', *Current Anthropology*, 47(2): 277–307.

Jenkins, Rob. 1999. *Democratic Politics and Economic Reform in India*. Cambridge: Cambridge University Press.

Jenkins, Rob. 2013. 'State Capitalism: Provincial Governance and Protest Politics', in Sudha Pai (ed.), *Handbook of Politics in Indian States: Regions, Parties and Economic Reforms*. New Delhi: Oxford University Press.

Jenkins, Rob and James Manor. 2015. 'Politics of Social Protection: The Mahatma Gandhi National Rural Employment Guarantee Act in Rajasthan and Madhya Pradesh', in Louise Tillin, Rajeshwari Deshpande and K.K. Kailash (eds.), *Politics of Welfare: Comparisons Across Indian States*, pp. 168–199, New Delhi: Oxford University Press.

Jenkins, Robert. 1995. 'Theorising the Politics of Economic Adjustment: Lessons from the Indian Case', *The Journal of Commonwealth and Comparative Politics*, 33(1): 1–24.

Kailash, K.K. and Madurika Rasaratnam. 2015. 'The Policy-shaping Capacity of States: Publicly Funded Health Insurance in Tamil Nadu and Kerala', in Louise Tillin, Rajeshwari Deshpande and K.K. Kailash (eds.), *Politics of Welfare: Across Indian States*, pp. 40–68. New Delhi: Oxford University Press.

Mitra, Subrata. 2007. 'Federalism's Success', in Sumit Ganguly, Larry Jay Diamond and Marc Plattner (eds.), *State of India's Democracy*. Baltimore, MD: Johns Hopkins University Press.

Nayar, Raj Baldev. 2013.'Globalization and State Disparities in India', in Sudha Pai (ed.), *Handbook of Politics in Indian States: Regions, Parties and Economic Reforms*. New Delhi: Oxford University Press.

Ohmae, Kenichi. (1995). *The End of the Nation State: The Rise of Regional Economies*. New York: Free Press.

Pai, Sudha, 2013. *Handbook of Politics in Indian States: Regions, Parties and Economic Reforms*. New Delhi: Oxford University Press.

Rudolph Lloyd and Susanne Hoeber Rudolph. 2001a. 'The Iconization of Chandrababu: Sharing Sovereignty in India's Federal Market Economy', *Economic and Political Weekly*, 36(18): 1541–1551.

Rudolph, Llyod and Susanne Hoeber Rudolph. 2001b. 'Redoing the Constitutional Design: From an Interventionist to a Regulatory State', in Atul Kohli (ed.), *The Success of India's Democracy*. Cambridge: Cambridge University Press.

Saez, Lawrence. 2002. *Federalism Without a Centre: The Impact of Political and Economic Reform on India's Federal System*. New Delhi: Sage Publications.

Saxena, Rekha. 2016. 'The Problems of Statehood in Indian Federalism: A Case for Territorial Pluralism', in Harihar Bhattacharyya and Lion König (eds.), *Globalisation and Governance in India: New Challenges to Society and Institutions*. London and New York: Routledge.

Singh, Ajay Kumar. 2003. 'Federalism and State Formation: An Appraisal of Indian Practice', in B.D. Dua and M.P. Singh (eds.), *Indian Federalism in the New Millennium*. New Delhi: Manohar Publishers.

Singh, M.P. and Rekha Saxena. 2013. *Federalizing India in the Age of Globalization*. New Delhi: Primus Books.

Sinha, Aseema. 2004. 'The Changing Political Economy of Federalism in India: A Historical Institutionalist Approach', *India Review*, 3(1): 25–63.

Sinha, Aseema. 2005. *The Regional Roots of Developmental Politics in India: A Divided Leviathan*. New Delhi: Oxford University Press.

Snyder, Richard. 2001. *Politics after Neoliberalism*. Cambridge: Cambridge University Press.

Tillin, Louise, Rajeshwari Deshpande and K.K. Kailash. 2015. 'Introduction: Comparing the Politics of Welfare across Indian States', in Louise Tillin, Rajeshwari Deshpande and K.K. Kailash (eds.), *Politics of Welfare: Comparisons Across Indian States*, pp. 1–39. New Delhi: Oxford University Press.

Tillin, Louise, Anupama Saxena and Yatindra Sisosdia. 2015.'Comparing the Politics of Food Subsidies in Chhattisgarh and Madhya Pradesh', in Louise Tillin, Rajeshwari Deshpande and K.K. Kailash (eds.), *Politics of Welfare: Comparisons Across Indian States*, pp. 102–134. New Delhi: Oxford University Press.

Verney, Douglas V. 2003.'From Quasi-Federation to Quasi-Confederacy? The Transformation of India's Party System', *Publius*, 33(4): 153–171.

Weiss, Linda. 1997. 'Globalization and the Myth of the Powerless State', *New Left Review*, 225(1).

Wheare, K.C. 1956. *Federal Government*. London: Oxford University Press.

10

EVALUATING THE INTEGRATED CHILD DEVELOPMENT SCHEME (ICDS) THROUGH INTERGOVERNMENTAL ENGAGEMENT

Jajpur district of Odisha

Sangita Dhal

Socioeconomic rights are enabling instruments of healthy democratic practices in every liberal society that lead to the empowerment of people in general and the marginalized and underprivileged sections of society in particular. These rights complement corresponding political and legal rights of citizens and thereby create conditions of good living and empowered citizenship. Health and education are the two basic criteria on which every society is judged, and accordingly placed in the ladder of the human development index. When we make a distinction between the developed and underdeveloped world, we try to figure out the levels of income and standard of life. One finds critical linkages between levels of development and quality of demographic character, which is defined in terms of quality of life, education and health. These are the significant objectives of economic discourse in any society. In other words, the degree of development achieved can be examined by the various parameters of the quality of life of the people, which is determined by the levels of health, education, vocation and overall inclusive growth.

The states across the world today are focusing on these aspects of development, wherein an integrated approach to develop the crucial social sector is being adopted by roping in the center and the states (federating units) and popular participation. It is being adopted with workable fiscal arrangements between them to fund such schemes and projects, which promote the interests of the underprivileged and vulnerable sections of the society through a robust service delivery mechanism. Studies have revealed that participation by stakeholders has a profound effect in quality and quantity of service delivery (Kannan 2014; Manor 1999). The benefits of decentralization provide for a more open, responsive and effective local government, greater public participation and improved quality of administration. Gerry Stoker argues for democratic decentralization and citizen-centric government which will erode the bureaucratic monopoly over the development processes, thus advocates a shift in the locus of power to those who matter at the grassroots (Stoker 1998; Venkatesu 2016).

Ever since India's independence, the priority sectors such as industry, agriculture and infrastructure building in the country relegated the social sectors like health and education to the background. The emphasis on the binary approach on either industry or agricultural development for overall economic growth was replaced by a holistic approach by integrating social sectors as well. This multidimensional approach to development laid emphasis on human resource development. It is in this context that the Integrated Child Development Scheme (ICDS) as a Centrally Sponsored Scheme was launched on 2 October 1975, which continues to be one of the largest and unique schemes in the world envisaging the holistic development of children under 6 years of age in the country. The nutritional status of children determines the development status of the country. The importance of nutrition outlined in the Convention on the Rights of the Child (1989) advises the states to take effective steps to reduce infant and child mortality and to address the perennial problem of malnutrition through enhancement of the nutritional status of children (Kumar and Banerjee 2015; Rachel Kumar 2002).

Context and background

This chapter establishes a positive correlation between three conceptual frameworks relating to decentralization, development and good governance characterized by restructured administrative organizations, reengineered work processes, strategic management, decentralization, delegated authority, participative vision and innovativeness among multiple stakeholders. The theoretical orientation of the research establishes the vital link between governance on the one hand and centrality of rights, liberty, equality and social justice for all its citizens to fully participate in the process of governance and development at all levels on the other. A robust economy and a vibrant democracy are dependent on how healthy the society is in terms of mother-child care, basic education, sanitation, nutrition and overall human resource development. The state through the Integrated Child Development Scheme (ICDS) project and the thousands of Anganwadi Centers (childcare centers) seeks to fulfill its social and constitutional responsibility by adopting a proactive approach toward the poor and underprivileged people of the region. ICDS is a social welfare program of the central government under the purview of the Ministry of Women and Child Development (MWCD) and seeks to integrate health, nutrition, sanitation, hygiene and education of children below 6 years of age, pregnant women, nursing mothers and adolescent girls. These services are provided in a cohesive manner at the Anganwadi Centers (AWCs), and each center operates through an Anganwadi worker (AWWs) and a helper. The center provides outreach services to low-income families in need of immunization, healthy food, clean water, clean toilets and a learning environment for infants, toddlers and preschool children and also provides similar services for expectant and nursing mothers. Located in rural and urban areas, the Anganwadi centers are the focal point for delivery of ICDS services and serve as the first point of contact for marginalized families experiencing nutrition and health issues in villages, towns and cities (ICDS Mission, MWCD 2012).

New model of development

India's development philosophy under the new economic policy in 1990s was redefined from economic growth and economic development to guaranteeing better quality of life to poor people by fulfilling their basic needs and fulfilling the minimum requirements of family. The focus was on developing a strong Indian democracy based on the foundations of a robust human development sector through intensive engagement with civil society (Singh 2016). In the mid-2000s, a new rights-based approach to social welfare policies was introduced across the Indian states that had a significant impact on the welfare scenario in contemporary India. Health care, food security and literacy missions began to receive priorities by the government because they played an instrumental role in facilitating other achievements and influenced welfare outcomes. Education and health policies were launched in 2004–2005 and 2011–2012 within the framework of the national goal toward greater socioeconomic empowerment of all communities hitherto marginalized and alienated in the development narratives of post-independent India; for example, the National Rural Health Mission (NRHM) began in 2005 (it later came to be known as National Health Mission) and the maternity security scheme titled Janani Suraksha Yojana (JSY), 2005, was implemented alongside the NRHM and Right to Education Act (RTE) in 2010. Substantial expenditure was also incurred in each of these Centrally Sponsored Schemes, with a demonstrative political will for ensuring allocation of funds to the schemes (Tillin, Deshpande and Kailash 2015). However, since the schemes are based on centrally sponsored initiatives and implemented at the level of the state government, they require a lot of coordination at the intergovernmental level, with clearly demarcated fiscal responsibilities on both the center and the state. This chapter delves into the micromanagement of the ICDS project in the district of Jajpur, situated in the state of Odisha and investigates the intergovernmental (center-state-district-block-village) cooperation and coordination or the lack of it while evaluating the performance of this scheme. It also analyzes the implication of ICDS on its stakeholders with the shifting of fiscal distribution in favor of either state or center. In other words, the study explores how the prospect of ICDS gets affected when there is a revision by the center with regard to the sharing of the financial burden with the state.

Research methodology

This chapter evaluates the impact of the ICDS program in the district of Jajpur in Odisha. The study was carried out in two phases, which included an initial in-depth pilot study followed by the field survey, which was conducted on the basis of initial feedback derived from the pilot study. The field investigation used both structured and unstructured questionnaires through personal interviews and focused group discussions. A pilot study was conducted in the December 2015 followed by field survey in May 2016. Since the present research required empirical investigation, the study adopted a social survey method for the collection of primary data through:

i Field investigation using participant observation
ii Interviews
iii Visit to relevant/concerned government offices.

Apart from gathering information about the various governmental agencies and non-state actors involved in the implementation of the ICDS, the other aim of the study was to gather opinions, suggestions and feedback across the sections of the population who have availed themselves of ICDS services. The case study method was used to assess the performance of ICDS from individual respondents. The total size of the sample was 100, out of which 60 respondents were ICDS beneficiaries (20 beneficiaries from each block; i.e., Bari, Dasarathpur and Jajpur in the Jajpur district); 30 respondents were from the ICDS official staff; five respondents were from the District Collectorate Office and the District Social Welfare Office; and five officials were from the Ministry of Women and Child Development (Figure 10.1).

At present there are 338 ICDS that are currently operational across the state of Odisha, with over 71,306 AWCs; 200 are rural (59 percent), 118 are tribal –(35 percent) and 20 are urban projects (6 percent) (Figure 10.2). In the sample district of Jajpur there are currently 11 ICDS projects that are functional through 2,603 Anganwadi centers covering over 280 village administrative units (village *panchayats*).

Odisha's ICDS intervention through the decentralized framework has been appreciated by the center and the judiciary. The efforts made by the Women and Child Development Ministry, Odisha, to decentralize the production of the Supplementary Nutrition Programme (SNP) involving community participation through a local self-help women's group is a best practice. The

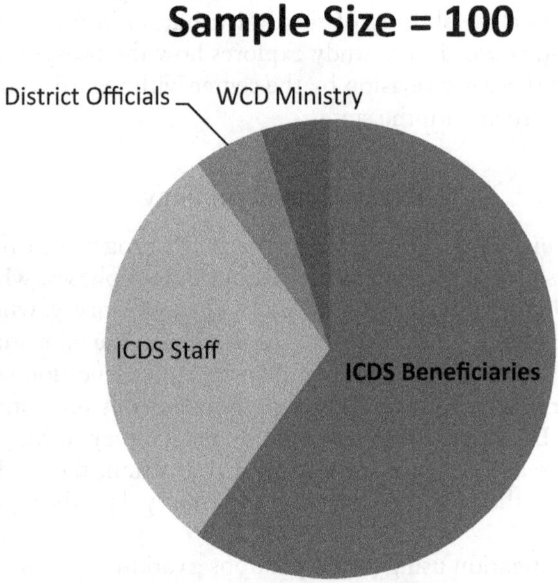

Figure 10.1 Sample Size and Types of Sample Respondents

Note: ICDS Staff refers to CDPOs, AWWs and AWHs; WCD = Women and Child Development Ministry

Source: Prepared by the author

central observers have recommended the model to be followed by other states. In another development, the Evaluation Report on ICDS by the then Planning Commission, now NITI Aayog, has also designated Odisha as a "High Performing State" (2013) with a frequency of delivery of SNP services of more than 80 percent.[1]

Decentralized structure for implementation of ICDS

Decentralization and development are significant themes in contemporary public administration. They are complementary and supplementary to each other, and therefore it is important to contextualize them in the present chapter. Decentralization is seen as a means of empowering people at the local level by involving them in the decision-making process that affects them. It is one of the best means of promoting efficiency at the grassroots level of local self-government (Maddick 1966). One of the main purposes of this research was to examine the modalities of the multilevel governance through a decentralized framework that impact the functioning of ICDS in Odisha. The process of decentralization aimed to address the poor service delivery of the developmental programs meant for the disadvantaged and the

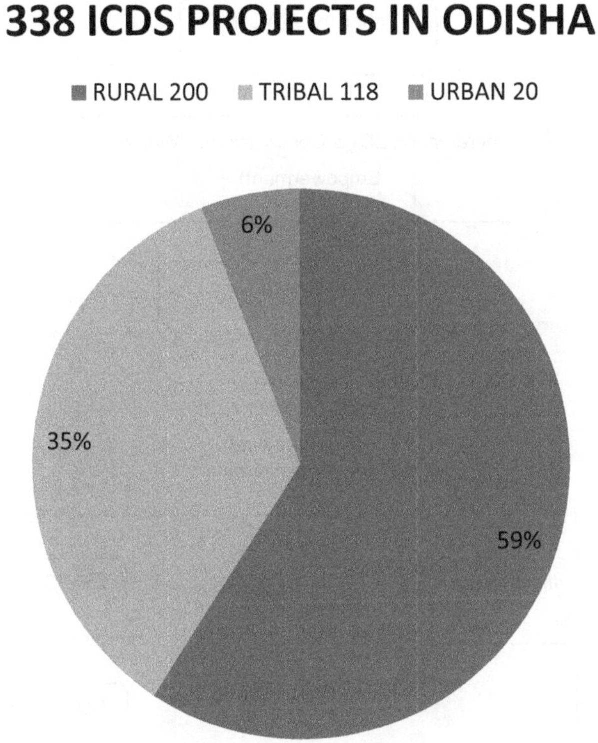

Figure 10.2 ICDS Projects in Rural, Tribal and Urban Areas of Odisha

Source: Prepared by the author

underprivileged sections of the society. The decentralized mechanisms and processes of the ICDS network highlight the need to conceptualize development both as a means as well as an end to achieve the larger objective of a just and equitable socioeconomic order. The intergovernmental coordination of the ICDS project is based on a decentralized framework wherein the central government has outlined the guidelines for monitoring the ICDS projects at the state level to ensure accountability and people's participation in preparing the local level plans (Figure 10.3). The decentralized framework of ICDS envisages the monitoring mechanisms at the various levels of the federal structure, which is outlined here.

Central level: The ICDS scheme is presently operating in the various states under the guidelines of the Central Ministry. It is strengthened and restructured through multi-sectoral programs in the states to address maternal and child malnutrition, sanitation, drinking water, primary health care, preschool education and food security.

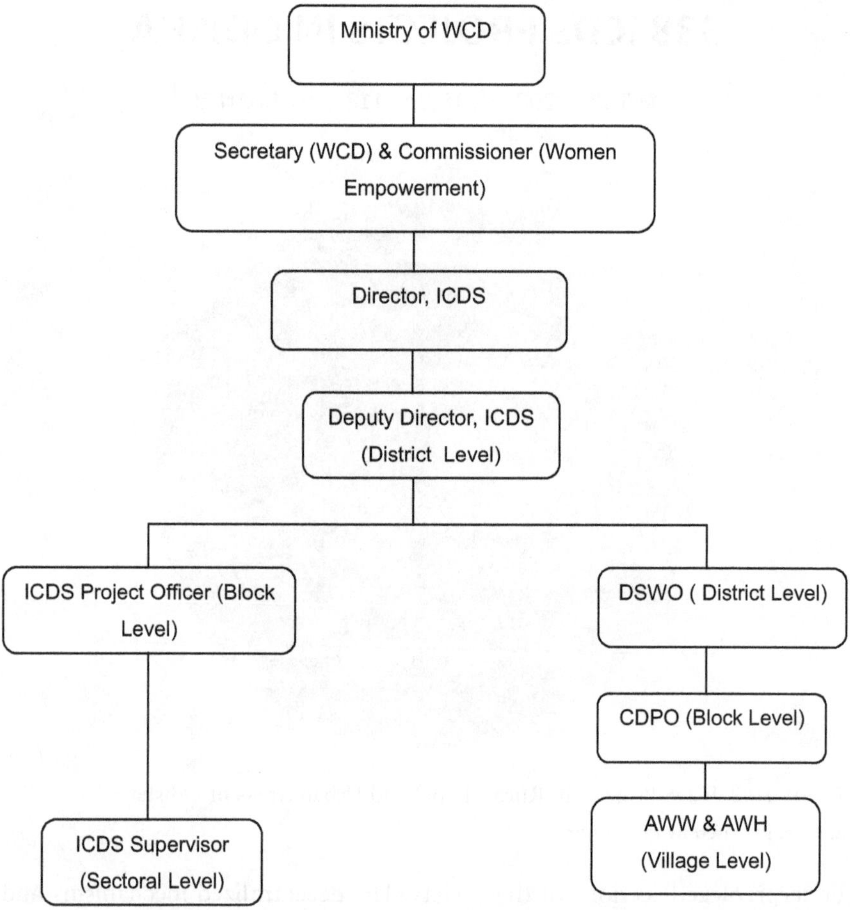

Figure 10.3 Decentralized Framework of ICDS Project

Source: Author

- CDPO–Child Development Programme Officer
- AWW–Anganwadi Worker
- AWH–Anganwadi Helper
- GPL–Gram Panchayat Level
- MWCD–Ministry of Women and Child Development
- DSWO–District Social Welfare Officer

State level: Various quantitative inputs are collected from the Child Development Project Officer (CDPO) monthly progress report (MPR) from various districts and submitted at the state level. CDPOs collect information on the number of beneficiaries for supplementary nutrition and preschool education and conduct regular field visits to AWCs along with other ICDS functionaries such as supervisors, welfare officer, etc. They submit the reports regarding implementation of the program on time and on prescribed guidelines.

District level: The district collector checks the monthly report of issues related to ICDS scheme. Subcollectors are the chairpersons of ICDS who provide support and guidance through reviews. At the village level, ward members, the Monitoring/Jaanch Committee and the Mothers' Committee are involved in implementation and monitoring of the various activities of ICDS.

The **District social welfare officers (DSWOs)** are entrusted with the job of implementing all the financial assistance schemes of the department and sanction financial assistance as per prescribed norms and rules under ICDS schemes. They also conduct surveys from time to time of eligible beneficiaries. They verify the work of supervisors – whether they are looking after AWWs – and guide them accordingly. DSWOs also supervise visits that are organized to the project offices and AWCs to see the actual implementation of the program.

Block level: The child development project officer (CDPO) is responsible for planning and implementation of the ICDS project at the block level and provide the Monthly Progress Report (MPR) to the district social welfare officer (DSWO). A CDPO is assisted by few supervisors who guide the AWWs with regard to the ICDS project. CDPO visits as many AWCs as possible (minimum 15 AWCs in a month) to assess the effective implementation of the ICDS program.

Village Level: At the grassroots level, delivery of various services to target groups is given at the Anganwadi Center (AWC). The basic job of the Anganwadi workers is very important, for they need to provide care to the newborn babies and ensure that all children below the age of 6 years are immunized. Along with the assistance of Anganwadi helpers, AWWs are expected to provide care for pregnant women and ensure that they are immunized against tetanus. In addition to this, they provide postnatal care to nursing mothers.

Collaborating agencies

There are a number of collaborating agencies that help in delivering the services and benefits of ICDS to achieve effective coordination of policy and implementation among the various departments to promote child development in a coordinated and meaningful manner.

Role of the Nongovernmental Organizations (NGOs)

The NGOs in the state of Odisha constitute a significant group of non-state actors who through their relentless efforts are creating conditions toward building a better society by helping the state government in successful implementation of various schemes designated for the benefits of women and children. The NGOs have made a substantial contribution by fulfilling the responsibility in delivering the best of services as desired by the government and adhering to the norms and standards devised for each individual scheme.

UNICEF and WHO

UNICEF funding to state government helps promote programs initiated to improve women's nutrition under the ICDS scheme, which helps to improve women's nutrition before, during and after pregnancy. UNICEF also conducts training programs for supervisors to prevent stunting and facility-based management of severely malnourished children. UNICEF provides technical assistance for effective functioning and provides weighing machines to the AWCs and growth monitoring registers and issues the new guidelines of WHO to the supervisors and AWWs. It also provides the vehicles for visiting AWCs, where fuel and driver is provided by the state government.

World Bank

The World Bank is providing support to Odisha by strengthening the Integrated Child Development Services (ICDS) projects and capacity building by emphasizing the health and nutrition of children less than 3 years of age and children between 3 and6 years of age, pregnant women, nursing mothers and adolescent girls. The World Bank assists the ICDS staff in formulating comprehensive plans and activities to improve nutrition outcomes for the beneficiaries. In a decentralized system, much will undoubtedly depend on the nature and functioning of the political institutions and processes at the local level. Elected local governments (*panchayats*) have also played a significant role in the implementation in Odisha since 1975.

District-wise allocation of AWCs in Odisha

There are a total number of 71,306 AWCs in the 30 districts of Odisha. The inland districts of Ganjam (4,968), Mayurbhanj (4,637) and Balasore (4,080) have the highest number of anganwadi center in the state, followed by the districts of Sundergarh (3,706), Koraput (3,264), Keonjhar (3,257) and Cuttack (3,205).

These centers serve as the grassroot development agencies for promoting child development, maternal and child health, nutrition and childcare support services. The sample district of Jajpur (which has been highlighted) has a total number of 2,603 AWCs, of which 2,578 are functioning as AWCs and 25 are operating as mini AWCs (Table 10.1). The mini AWCs have been

Table 10.1 List of Districts in Odisha with Total Anganwadi Centers, May 2016

Serial no.	*Name of the District*	*No. of AWCs*	*No. of Mini AWCs*	*Total AWCs*
1.	ANGUL	1,614	32	1,646
2.	BOLANGIR	2,286	157	2,443
3.	BALASORE	3,630	450	4,080
4.	BARGARH	2,670	229	2,899
5.	BHADRAK	2,166	269	2,435
6.	BOUDH	650	67	717
7.	CUTTACK	2,879	326	3,205
8.	DEOGRAH	572	216	788
9.	DHENKANAL	1,856	324	2,180
10.	GAJAPATI	1,442	924	2,366
11.	GANJAM	4,937	31	4,968
12.	JAGATSINGHPUR	1,647	193	1,840
13.	JAJPUR	2,578	25	2,603
14.	JHARSUGADA	786	171	957
15.	KALAHANDI	1,830	408	2,238
16.	KANDHAMAL	1,580	522	2,102
17.	KENDRAPADA	1,883	264	2,147
18.	KEONJHAR	2,512	745	3,257
19.	KHORDHA	2,373	139	2,512
20.	KORAPUT	2,723	541	3,264
21.	MALKANGIRI	1,020	230	1,250
22.	MAYURBHANJ	3,465	1,172	4,637
23.	NAWAPARA	1,216	140	1,356
24.	NOURANGPUR	2,038	169	2,207
25.	NAYAGARH	1,142	402	1,544
26.	PURI	2,560	39	2,599
27.	RAYAGADA	1,397	550	1,947
28.	SAMBALPUR	1,466	376	1,842
29.	SONEPUR	1,250	321	1,571
30.	SUNDERGARH	2,922	784	3,706
GRAND TOTAL		61,090	10,216	71,306

Source: Women and Child Development Department, Government of Odisha, May 2016

provided by the government to cater to the needs of the scheduled caste and scheduled tribe population, who often reside in difficult geographical terrain, and to spread the level of awareness about the different ICDS services among the historically disadvantaged groups across the district so that they can effectively utilize the ICDS services for the benefit of the women and children.

Federal fiscal transfer

Largely, the ICDS scheme receives funds from the central government, and partly from the state budget. Odisha receives central grants for Early

Childhood Care and Development (ECCD) programs under the WCD Department under two separate heading, the Child Development Grant and ICDS. It encompasses the elements of care, health, nutrition, play and early learning within a protective and enabling environment. It is observed that for financial year 2012–2013 no fund was released under ICDS by the central government till December 2012. All the funds were transferred to the state in January and February 2013. This untimely and skewed pattern of fund release by the central government ultimately affects the functioning and quality of the program. As per the census 2011, the state of Odisha has 5.03 lakh of population in the age group of 0–6 years; despite constituting 12 percent the state's population (0–6 years age group), the allocation for ECCD programs in 2012–2013 was merely 3.14 percent of the total state budget and 0.63 percent of the Gross State Domestic Product (GSDP). This has failed to provide a better quality of life to the young children of the state. In recent times, the center, in order to decrease the current dependence of the state on the center for federal fiscal transfer, has reduced the allocation of ICDS budget to a 60:40 ratio from the earlier 75:25 ratio, a move which has been opposed from various quarters.

ICDS is a Centrally Sponsored Scheme implemented through the state government and union territory administrative structures. Prior to 2005–2006, the center provided the total financial assistance (inputs other than supplementary nutrition were being provided by the center). This policy was reviewed by the center, and 10 percent of the financial burden of the ICDS expenditure was shifted to the states on supplementary nutrition program. From the financial year 2009–2010, the Government of India had modified the funding pattern of ICDS between center and states for all other components of ICDS as 90:10 (in contrast to 100 percent central assistance earlier). This was further revised to a 75:25 ratio in April 2011, which marks a steady rise in the financial burden of the state to carry out its responsibilities toward ICDS in the state (refer Figure 10.4).

The scenario in the social sector expenditure was altered drastically by the present NDA government, which revised the share between center and the state to a 60:40 ratio in the financial year 2015–2016 (Figure 10.5). The social sector indicators can be improved not only by investments but also by investing in the allied sectors, which are complementary to the social sectors. In some areas with welfare implications, such as health, sanitation, family welfare, preschool education and literacy, the incremental spending projected for 2016–2017 is not much above the difference between the revised estimate for 2015–2016 and the actual estimate for 2014–2015. While health insurance schemes for the poor are welcome, they do not make up for the gross inadequacy of spending in related areas. In the case of women and child development, where spending fell by Rs.1,188 crore in 2015–2016 compared to 2014–2015, the budgeted increase in spending in 2016–2017 is just Rs.56.23 crore.

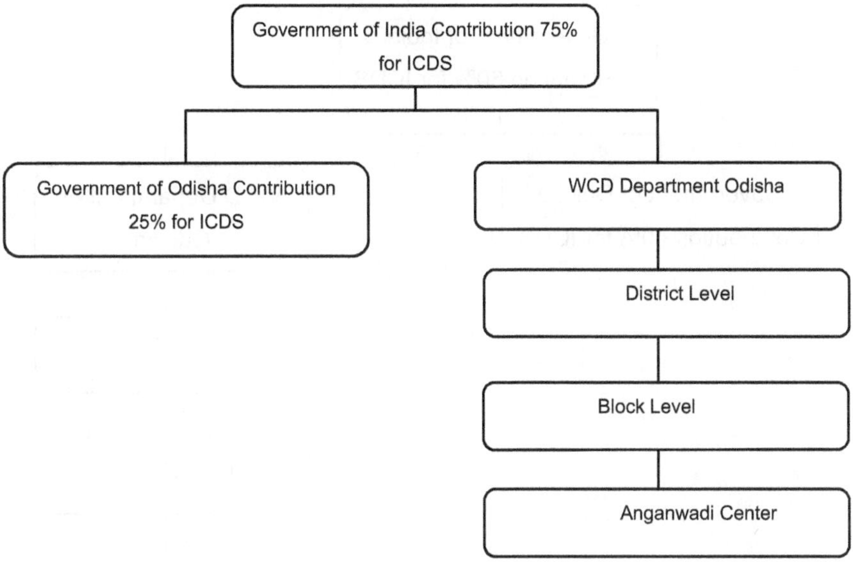

Figure 10.4 Finance Contribution of the Central Government and State Government, 2011–2014

Source: Author

As a result, spending on the important program of ICDS was sadly reduced from Rs.16,415 crore in 2014–2015 to Rs.13,636 crore in 2015–2016 and will receive only Rs.15,873 crore.in 2016–2017 (Chandra Shekhar 2016). This cut in the budget has also resulted in the reduction of a budgetary allocation for preschool education, which is one of the key components of the ICDS program. It includes provision for the Pre School Education Kit (PSEK), Toy Bank and Activity Book for the children and celebration of Early Childhood Education Day for community awareness. For this purpose, the AWCs and mini AWCs receive Rs. 2,500/year and Rs. 1,250/ year, respectively. This reduction of central funds has a tremendous negative impact on the functioning of the ICDS scheme, which resulted in the nonpayment of salaries of the ICDS staff in many blocks and also supplies of uniforms to the staff and children. (Maity 2016)

However, the recent decision by the central government to reduce the budgetary allocation for ICDS will have a negative impact on further strengthening the implementation process of the project (Figure 10.6). The budgetary allocation for the ICDS scheme in the financial year 2015–2016 was reduced to almost 50 percent as compared to last two financial years. This financial year, the allocation is just Rs. 8,335.7 crore as the central share, whereas the budgetary allocation amount for 2013–2014 and 2014–2015 for the ICDS scheme was Rs. 16,312 crore and Rs. 16,561 crore, respectively.[2]

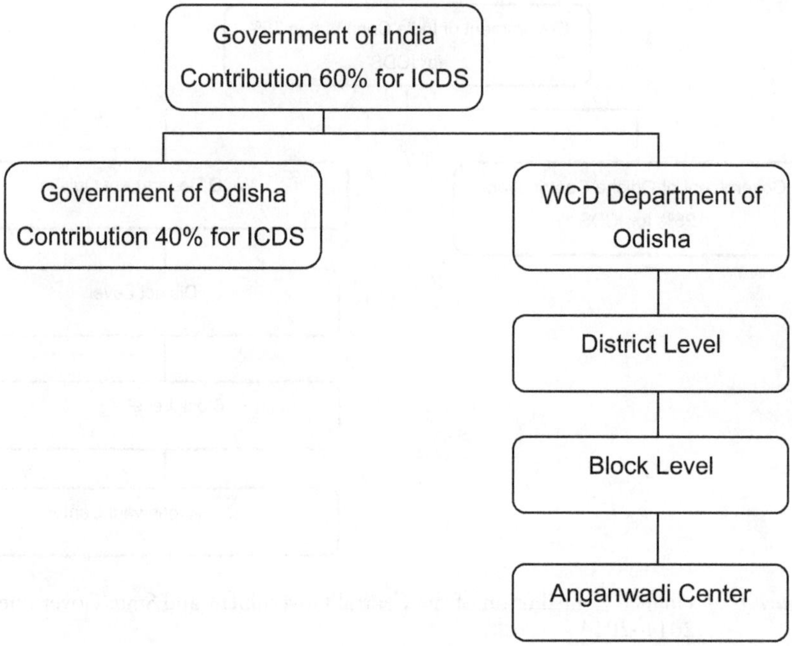

Figure 10.5 Finance Contribution of the Central Government and State Government, 2015–2016

Source: Author

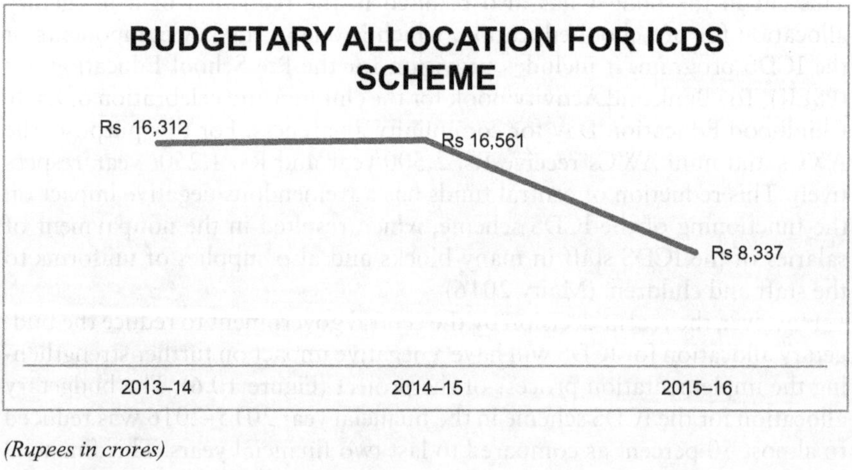

(Rupees in crores)

Figure 10.6 Budgetary Allocation for ICDS Scheme

Source: Ministry of Women and Child Development, Government of India, 2015

The recent decision leading to drastic reductions in ICDS budget has negatively impacted the ongoing strengthening and restructuring of ICDS schemes that had already started a series of programmatic, management and institutional reforms, including placing ICDS in mission mode project as envisioned and approved under the 12th Five-Year Plan period. Ironically, the draft concept note of the "Smart City Scheme" of the Government of India in which Bhubaneswar, the capital city of the state, has earned the first position does not incorporate the ICDS scheme. The draft proposal on the smart city underlined the importance of health, sanitation and social infrastructure; however, there was no reference of ICDS services or addressing undernutrition among the urban poor and slum settlements.[3] Thus, there is a pertinent need to strengthen the decentralized planning processes through community participation for effective utilization of the ICDS funds. The officials in the women and child development department highlighted that in many instances the districts do not have a uniform reporting format for budgets and financial statements. Further, the state government maintains that the center should support with required budgetary allocations the major activities undertaken to restructure and strengthen ICDS in a meaningful and effective manner.

Critical components of ICDS

The existence of poor health conditions and inadequate health services in India are responsible for incidents of infant mortality, maternal mortality, casualties in the sterilization camps and vaccination drives and midday meal scams due to lack of proper facilities. These are reflections of dysfunctional and unaccountable health services, which raise serious doubts about the ability of the state as a service provider to the common man. The various programs for good health and literacy promoted by the government provide holistic solutions toward ensuring an enhanced economic status for the common man. For this purpose, the ICDS in its programs includes critical services like medical check-ups, immunization, growth promotion, supplementary feeding, referral services, early childhood care, preschool education, and nutrition and health education, which are delivered through the local primary health centers and the Anganwadi Centers (AWCs) (https://www.cysd.org/wp-content/uploads/2019/11/Perception-Study-PHC.pdf).

The Government of Odisha took many necessary steps in April 2011 to reform and revitalize the ICDS system through decentralization. In the new system, for greater involvement of the community in actual implementation, Anganwadi workers (AWWs) were directed to obtain all the materials except rice and wheat in order to reduce chances of pilferage during transmission.[4] The involvement of contractors in ICDS projects is sought to be done away with because of their commercial interest, which forces them to compromise on the quality of food supplies. Accountability and transparency are the two other benchmarks that the Supreme Court has set for these ICDS projects to maintain, while pursuing the objective of integrated child development. Under the honorable court's directions, the state shall take steps to ensure the quality of food items with requisite protein and calorie intake of children, besides the priority given to the local palate and choice of food. This renewed focus shows a transformative shift toward ensuring entitlements to the target groups by an inclusive and participative mechanism, involving women's self-help groups (WSHGs) and checks and balances through a proper monitoring system.

Empirical findings

The general public perception among the respondents who were interviewed has been favorable with regard to the positive impact of ICDS on the health and nutritional status of children that helped in imparting a playful method of education and reducing infant mortality. However, the shortages of funds and dearth of functionaries affected the effective functioning of the ICDS in the various districts of Odisha. The impression among the respondents is that for efficient delivery of services, the government should increase its focus on a robust public delivery system. The ICDS staff suggested simple administrative procedures and reducing the number of agencies involved in the process of implementation of the ICDS projects.

Monitoring mechanism of AWCs

Sector-level review based on population size is conducted for the six package programs of ICDS relating to health check-up, immunization, nutrition, referral services, supplementary feeding and preschool education. The CDPO has a 20-day tour program every month and submits the monthly progress report to DSWO.

Functioning of the AWCs

The survey revealed that the Anganwadi centers function 25 days in a month and provide children below 6 years of age with hot cooked meals including rice. For 20 days in a month, the AWC functions and feeds the children for Rs 150 per child for a hot cooked meal including rice. Eight eggs and two packets of protein-rich cereal powder are supplied every month to every beneficiary of the AWC. There are well-established mechanisms available for monitoring the service delivery and public grievance resolution: project level meetings, sector level meetings and village level meetings in addition to official meetings are held as per guidelines provided by the government of Odisha from time to time. The CDPO provides the monthly progress report to the DSWO.

Out of the total 71,134 AWCs existing in Odisha, only 16,000 AWCs had their own concrete building till December 2013. In 2013–2014, an amount of Rs. 200.00 crore was allocated for construction of 4,000 new AWC buildings. Presently, 37,134 AWCs in the state are running in temporary accommodations such as in a community hall, youth club, rented house, etc. As per the revised guideline of the ICDS mission there is provision of Rs 700 and Rs 3,000 for rent of AWCs per month in rural areas and urban areas, respectively, which these centers are using to run the services (Women and Child Development Department, Odisha, 2015).

In the sample area of Jajpur Block of Jajpur district, there are 307 AWCs, including Jajpur Municipality, which has 40 Anganwadi Centers that are managed by eight supervisors. The field staff visits are conducted to supervise and monitor 15 AWCs per month to attend project meetings and health meetings to check registers, records and preschool data. The funds allocated for hot cooked meal and morning snacks are transferred to the joint accounts of AWW and the ward member of the concerned village.

These data revealed that out of the total 2,603 AWCs, 648 (24.89 percent) of AWCs operate from their own premises, whereas a majority of AWCs, 1,965 (75.10 percent), operate from other rented premises (Table 10.2). The evaluation of the performance of AWCs across the district of Jajpur can be based on four major parameters such as kitchens, electricity connection, toilet provision and drinking water facilities, which can be further classified under two categories such as AWCs operating from their own premises and AWCs functioning from other rented places.

It has been found that out of 11 blocks in the district of Jajpur where AWCs are operational, there are fewer AWCs operating from their own buildings as compared to the large numbers of AWCs functioning from rented premises. For example, Sukinda block has the maximum number of AWCs (106) operating from their own buildings followed by Jajpur block with 81 and Danagarhi with 80 AWCs operating from their own premises. AWCs operating in other premises is highest in Jajpur block (226), followed by Sukinda (148) and Danagarhi (82). Out of 136 AWCs functioning in the Badachana block, it has only 22 AWCs operating from own buildings and also records the least number of AWCs operating from other buildings (114) as compared to the other ten blocks. Dharamshala block has the highest number of AWCs (252) operating from rented premises followed by Jajpur (226) and Dasarathpur (221).

Kitchen facilities

Sukinda block has maximum functional kitchens operating from own premises (106), whereas Badachana (15) and Binjharpur (15) have the least number of functional kitchens operating from their own premises. The second largest number of kitchens operating from their own premises is Danagarhi (80).

Table 10.2 AWC Infrastructure for the Month of June 2016 of Jajpur District

Sl no	*Name of Block*	*No of AWCs/ Mini AWCs functioning in own building*	*No. of AWCs having kitchen*	*No. of AWCs having toilets*		*No. of AWC having drinking water facilities*		*No of AWCs/ mini AWCs functioning from rented premises*	*No. of AWCs having kitchen*	*No. of AWCs having toilets*		*No. of AWC having drinking water facilities*	
				Functional	*Non functional*	*Tube well*	*Piped water supply*			*Functional*	*Dysfunctional*	*Tube well*	*Piped water supply*
1	2	3	4	5	6	7	8	10	11	12	13	14	15
1	Badachana-R	22	15	0	0	4	0	114	0	17	0	37	0
2	Badachana	34	37	34	3	28	0	122	37	34	3	28	0
3	Bari	42	21	21	0	19	0	197	0	0	0	207	0
4	Binjharpur	37	15	7	2	36	1	247	0	20	227	247	5
5	Danagadi	80	80	68	12	76	2	82	0	0	0	79	0
6	Dasarathpur	62	45	45	0	49	0	221	164	164	0	164	0
7	Dharmasala	57	41	41	0	38	3	252	0	0	0	230	15
8	Jajpur	81	71	71	0	81	0	226	0	0	0	102	0
9	Korei	73	73	8	11	10	0	154	0	0	0	100	54
10	Rasulpur	54	41	41	0	41	0	192	0	0	0	189	0
11	Sukinda	106	106	55	51	94	6	148	0	0	0	120	10
Total		651	549	396	85	483	20	1,965	212	247	243	1,517	99

Source: District Social Welfare Office, Jajpur

The highest number of kitchens in AWCs operating from other premises is Dasarathpur block (164), followed by Badachana (37). The study also reveals that there are no functional kitchens in most of the AWCs that are operating from other buildings, and in most instances, the Anganwadi helpers cook food from their homes and bring it to the Anganwadi centers.

Water supply

AWCs operating in all the 11 blocks from both own building and rented premises have drinking water supplies made available through tube wells and pipes. AWCs in Sukinda block having their own building have the maximum number of tube well water supplies (94) followed by 81 and 76 in Jajpur and Danagadi blocks, respectively. Badchana-R has the least number of tube well water supply (4) and Sukinda has the most piped water supply (6), followed by Dharamshala block (3) and Danagarhi (2). The rest of the blocks, except Binjharpur (1), have no piped water supply in the permanent buildings of AWCs.

The AWCs operating from rented premises have both tube well as well as piped water supplies. Binjharpur records the highest number of tube wells (247). The Dharamsala block has 230 tube wells in the AWCs, followed by 207 tube wells in the Bari block. Korei records the highest number of piped water supply (54) in the AWCs operating from rented premises, followed by 15 in Dharamsala and 10 piped water supplies in the Sukinda block.

Electricity

It is noteworthy to mention that Dharamsala block has the most (37) AWCs operating from own buildings, where there is electricity facility. However, in all other blocks of the district except Korei and Badachana, there are 62 AWCs with electricity facilities. The AWCs operating from other buildings in the blocks of Dasarathpur have the highest number of AWCs with electricity facilities (164), followed by Dharamsala (70), Korei (10), Badachana and Binjharpur, with two each.

Toilet facilities

It has been noticed that in all the 11 blocks of Jajpur district, AWCs operating from both own and rented premises toilets were found in both functional and dysfunctional conditions. Jajpur block has the highest number of functional toilets operating from own buildings of the AWCs, followed by 68 and 55 functional toilets in Danagadi and Sukinda, respectively. Sukinda records the highest number of dysfunctional toilets (55), followed by 12 and 11 in Danagadi and Korei.

As far as AWCs operating from other buildings are concerned, Sukinda block has the maximum number of functional toilets (235) and also the highest number of dysfunctional toilets (230), followed by Binjharpur, with 227, and Badachana having only three dysfunctional toilets. Thus Dasarathpur appears to have struck a balance between kitchen and functional toilets, while the rest of the blocks in Jajpur district having AWCs operating from other premises have few dysfunctional toilets.

Though there has been a substantial presence of ICDS in the rural areas, improving the infrastructural facilities in the AWCs has been a major challenge for the CDPOs. In most cases, the ICDS staffs has to overcome host of issues pertaining to the nonpayment of salaries, lack of trained staff, shortage of supply of food grains, lack of proper drinking water and lack of toilet facilities.

The positive outcome

The success percentage of institutional deliveries among pregnant women in the rural areas has been heartening, due to the consistent efforts made by the AWW and AWH in response to the cash incentives they receive. Although the quality of maternal care remains a major concern, increased usage of the government services points to the significant success of the program in rural areas.

Table 10.3 reveals that the incentive paid to each Anganwadi worker per institutional delivery is Rs 200 and each Anganwadi helper gets an incentive of Rs 100. A total amount of Rs 15,759,800 has been provided under the incentive category to the AWWs from 2013–June 2016 and the total amount paid to the AWHs has been 7,647,200. This data reveals that the ICDS local officials are in touch with the various households in their respective areas to report and assist cases of pregnancy and delivery. The demand for increase in the cash incentive and honorarium for the AWWs and AWHs was conveyed to the researcher during the course of the survey.

Table 10.3 Year-wise AWW & AWH* Incentive Coverage in Jajpur District

Year	*Aww Incentive Paid (Rs 200)*	*Expenditure*	*Awh Incentive Paid (Rs 100)*	*Expenditure*
2013–14	19,527	3,905,400	18,400	1,840,000
2014–15	15,646	3,129,200	14,576	1,457,600
2015–16	43,626	8,725,200	43,496	4,349,600
TOTAL	**78,799**	**15,759,800**	**76,472**	**7,647,200**

** AWW- Anganwadi Worker*
AWH- Anganwadi Helper

Source: District Social Welfare Office, Jajpur, Odisha, June 2017

Political factor

One of the concerns of the present research study was to examine the influence of the regional political environment and its impact on the formulation and implementation of the ICDS project. The existence of politico-ideological homogeneity in the district as far as the electoral representatives are concerned leads to smooth consensus building because they belong to the same ruling party in the state. This gets translated into unanimous decision making. In others words, the 11 MLAs of the dominant regional political party, Biju Janata Dal (BJD), in the district, who represent the state government as well, decide on issues and challenges of implementation of ICDS in a more cohesive and substantial manner in the absence of political opposition from the rival parties, who often are seen as stumbling blocks for delay in the implementation process. However, political interference is often experienced in filling of Anganwadi posts of program officers, supervisors, workers and helpers that have been vacant for a long time. The absence of the staff has affected the functioning of centers in the various villages across the district. Efforts should be made to make the organizations more independent so that they are able to function in an autonomous and meaningful manner.

Addressing the challenges

Despite the honest and sincere endeavors of the state government and the center, the ICDS projects are often confronted with myriad challenges. One of the pressing problems that require urgent attention of the center is the condition of the AWCs where the small children are provided preschool education and nutrition. Sometimes the situation in the Anganwadi centers depicts a disheartening scenario with unhygienic surroundings. Poor infrastructure, lack of resources, overburdened staff, pilferage and lax enforcement make effective implementation of the ICDS project challenging in the district. Constraints of space and sanitation, lack of training of ICDS staff and inadequate provision for childcare plague the functioning of ICDS.

Probably this explains why within a span of three months in July 2016, 19 children died due to malnutrition among the indigenous and primitive Juang tribes in Nagada, a tribal village in the Sukinda block of the Jajpur district, which is located in the mineral-rich area stretching across Jajpur, Keonjhar and Sundergarh districts. Despite integrated action plan programs, there are no schools, no Anganwadi centers, no health sub-center and no fair-price shops or nonmotorable road. After 21 July 2016, three mini Anganwadi centers for the 85 Juang tribe families have been sanctioned by the district administration, under which the tribal families are now covered under the food security programs.

Addressing the health and nutrition of poor children is both a right and an equity issue. There is a need to relook at the ICDS and see how it can be effectively used to tackle nutrition issues in poor children. In 2013, the Comptroller and Auditor General (CAG) Report had criticized the Women and Child Development Ministry that is piloting the project for diverting the ICDS funds; a few state governments also had diverted Rs 57.82 crores from the ICDS budget to other activities.[5] Field report suggests that students like the hot cooked meal and eggs they get in the AWCs. Kids look forward to the *Annaprasana Diwas* (first rice eating ceremony), wherein a steel bowl and spoon are given to 6-month-old children to cat sweet porridge (*kheer*), so all the children at the AWCs are fed with this dessert on that day.

The beneficiaries urged the researcher to suggest measures to increase transparency to curb corruption, as it is a major hurdle for effective rural development and poverty alleviation. The study also found that often, beneficiaries complained about the irregular facilities provided in the AWCs and lack of basic amenities like drinking water facilities and toilets. Insufficient money allocated to each child per day, that is, Rs 4.52 toward a hot cooked meal and Rs 1 for morning snacks, is considered to be a major constraint to maintain both quality and quantity of meal to the children. Eggs, which are a major source of protein, are provided on only on Wednesday and Saturday, which the children are unhappy about and prefer eggs every day.

It is quite ironic that in the times of *Swachha Bharat Abhiyan* (Clean India Campaign) and *Beti Bachao Beti Padho Abhiyan* (Save Girl Child and Educate Girl Child Campaign) one comes across disheartening scenarios in these centers. Hot cooked meals are served to these small kids at a heavy price, because not only is the food cooked in unhygienic conditions infested with insects and dirt, the smoke the emanates from the cooking process travels into the small rooms. The vulnerable kids spend their day inhaling the smoke, unmindful of its poisonous effect. The present "SMART" (simple, moral, accountable, responsive and transparent) government must take cognizance of this pressing problem and provide LPG cylinders for cooking purposes to these anganwadi centers throughout the state of Odisha to cook healthy food under healthy conditions for the small kids. These observations reflect on the necessity for further improving the delivery system envisaged in the ICDS project through a thoughtful and meaningful approach involving all the stakeholders, which will have a greater potential for enhancing health and educational outcomes.

The way forward

In order to make the ICDS implementation successful, the government needs to focus on a few things, such as efficient delivery of services and public good, in a time-bound manner. Effective, efficient and timely delivery of services ensures the success of the ICDS project; otherwise, despite investment and funding, the government will not achieve its goals of a well-performing state in terms of ICDS outcome. Second, the government needs to simplify the procedure of service delivery and minimize the number of agencies involved in this process. The simpler the process, the greater is the likelihood of the success of the scheme. It is often seen that a multiplicity of agencies involved in the service delivery mechanism leads to an exercise of dual authority and delay. Third, the transparency quotient is considered to be the bedrock of a robust public delivery system, which the government should ensure to build a democratic, participative and inclusive model of good governance. The government today is expected to devise new methods, formulate new ideas and follow innovative practices to bring about this transparency and objectivity in administration. For a comprehensive analysis of ICDS, policy makers should identify and explore solutions for the beneficial outcome of ICDS, introduce more training programs for the ICDS staff and adopt the best practices, which can support the ICDS in an effective and meaningful manner. During the fieldwork, it was found that the level of knowledge among the AWWs was helpful in raising awareness about the importance of ICDS among the scheduled caste and scheduled tribe mothers, thereby raising potential demand for ICDS services in their areas. The few measures to recommend improving the working of ICDS:

- Availability of basic infrastructure like Anganwadi centers, water, electricity and LPG cylinders, etc.
- Accessibility of Anganwadi centers for small children at convenient locations.
- Quality of food and trained Anganwadi workers
- Suitable policies must be adopted to address the nutritional and health concerns of the poor children.
- Community participation is very important, so steps should be taken to encourage more community participation among the various stakeholders
- There is a pressing need to review the ICDS by the central management; techniques and institutional reforms are necessary to strengthen the intergovernmental network for the better functioning of the scheme.
- Center needs to continue with a proactive role and should not relinquish its responsibility toward funding and indirectly managing the ICDS.

A culture of good governance is therefore essential to rebuild the lost confidence of people, which will ensure not only quality services but also improve accountability, transparency and timely delivery of services.

Conclusion

As noted earlier, the period between 2004–2005 and 2011–2012, which witnessed a shift toward a rights-based approach to development with the introduction of several centrally sponsored programs, generated a lot of excitement and enthusiasm across the country. It is observed that most of government's initiatives based on good intent and purpose get defeated due to inefficient administrative mechanism and inept handling of public needs. This is partly due to archaic methods of governance and partly due to apathy and indifference of lower level bureaucracy. It is also the lack of transparency, responsiveness and accountability in the functioning of a multitude of governmental institutions and public services delivery systems that affect the outcome of every governmental initiative. Therefore, it is desirable to achieve a paradigm shift in the whole approach to a people-centric, bottom-up strategy so that schemes like ICDS can be successful. It is believed that the larger the participation, the greater the prospect of transparency and accountability in governance. The entire world is now confronted with this challenge of redefining the goals of economic development and rethinking democratic methods and processes to make governance more equitable and inclusive. The comprehensive and integrated early childhood program is regarded as an investment in the future economic and social progress of the country. Thus, ICDS, which is a flagship social welfare program sponsored by the center, needs a greater degree of coordination and cooperation between the center and the states so that the program can realize its full potential of ensuring timely health care and nutrition services to the poor and needy children.

Notes

1 'Odisha ranked as High Performing State in ICDS Interventions' www.orissadiary.com/CurrentNews.asp?id=38839#sthash.Yw2un9Od.dpuf, Jan 14,2013, accessed on 22 June 2016.
2 Press Information Bureau, Ministry of Women and Child Development reply to the Rajya Sabha on the issue of budgetary allocation toward ICDS project, 19 March 2015.
3 Draft Concept Note on Smart City Scheme by Ministry of Urban Development, Government of India dated 3 December 2014.
4 *The Hindustan Times* (New Delhi) 'Failing Our Children', 29 June 2016.
5 The ICDS Projects are now being implemented in the various districts/blocks and villages of Odisha on the basis of the Supreme Court guidelines, which give utmost importance to the role of local communities and institutions in service delivery.

References

Center for Youth and Social Development, Bhubaneswar, Odisha. 'Issues and Challenges of Primary Healthcare Facility', 2019. https://www.cysd.org/wp-content/uploads/2019/11/Perception-Study-PHC.pdf, Center for Youth and Social Development, Bhubaneswar, Odisha (accessed on 22/09/2020).

Chandra Shekhar, C.P. 2016. 'The Human Face of the Budget 2016–2017', *The Hindu*, March 1, 2016.

Kannan, K.P. 2014. *Interrogating Inclusive Growth: Poverty and Inequality in India*. London: Routledge Publishers.

Kumar, Rachel. 2002. 'Gender in Reproductive and Child Health Policy', *Economic and Political Weekly*, Vol. 38, No. (32).

Kumar, Sanjeev and Sainath Banerjee. 2015. 'Integrated Child Development Services (ICDS) Programme in the Context of Urban Poor and Slum Dwellers in India: Exploring Challenges and Opportunities', *Indian Journal of Public Administration*, Vol. LXI, No. 1, January–March 2015, pp. 94–113.

Maddick, Henry. 1966. *Democracy, Decentralization and Development*. Mumbai: Asia Publishing House.

Bipasha Maity, Bipasha. 2016. 'Interstate Differences in the Performance of Anganwadi Centers Under the ICDS Scheme', *Economic and Political Weekly*, Vol. LI, No. 51, December.

Manor, J. 1999. *The Political Economy of Democratic Decentralisation*. Washington, DC: World Bank.

Ministry of Women and Child Development (MWCD). 2012. *ICDS Mission: The Broad Framework for Implementation MWCD*, New Delhi.

Ministry of Women and Child Development (MWCD). 2014. *State Wise Number of Sanctioned, Operational ICDS Projects & AWCs, MWCD*, Government of India, New Delhi.

MWCD and NIPCCD. 2012. *Draft Report on Strengthening Maternal and Child Care*. Nutrition and Health Services in Urban Settings, July 18–19, 2012, National Institute of Public Cooperation and Child Development (NIPCCD), MWCD, New Delhi.

Singh, Satyajit. 2016. *The Local in Governance: Politics, Decentralization and Environment*. New Delhi: Oxford University Press.

Stoker, Gerry. 1998. 'Governance as Theory: Five Propositions', *International Social Science Journal*, Vol. 50, No. 155, pp. 17–28.

Tillin, Louise, Rajeshwari Deshpande and K. K. Kailash. 2015. *Politics of Welfare: Comparisons Across Indian States*. Oxford: Oxford University Press.

Venkatesu, E. 2016. *Democratic Decentralization in India: Experiences, Issues and Challenges*. New York: Routledge Publishers.

Women and Child Development Department, Government of Odisha, 2015 (https://finance.odisha.gov.in/Budgets/Outcome_Budget-2015-16/WCD_outcome_budget_2015-16.pdf, accessed on 10.01. 2018).

11

ROLE OF STREET-LEVEL BUREAUCRATS IN MGNREGS IMPLEMENTATION

Jharkhand

Rahul Mukand

"Inclusive Growth" was one of the themes for the 11th five-year plan, from the year of 2007 to 2012, that former Prime Minister Manmohan Singh elucidated in his lectures in the government-led think tank NITI Aayog (erstwhile Planning Commission of India).[1] The central vision of the 11th five-year plan was to upgrade India's developmental process, with a focus on improvement in the human development index for India. This was only possible if India adopted an inclusive growth model that incorporated all segments of Indian society. This theme portrays the commitment of the Indian federal state to provide development-related intervention in rural poverty. However, the ground reality contradicts the tall claims made by the Indian state at the top level.

If we study the budgetary allocation in the general budget in the years 2000 to 2004, the government's expenditure on social welfare schemes were reduced. Political economists (Corbridge and Harriss, 2013) argued Indian state revenue collections have gone up over the years, but that has not led to an increase in budgetary allocations for social programs. This brings us to question the real intentions of the Indian state in MGNREGS implementation. Does the Indian state work efficiently at a local level? To seek an answer to this question, fieldwork was conducted in five villages of Jharkhand in December 2014, and the results are presented in this chapter.

The constitutional perspective justifies the real intentions of the Indian state to provide necessary living and working conditions that can improve the plight of the poor in India. Article 21 in the Constitution of India provides entitlements for the rural poor; in terms of "Protection of life and personal liberty, no person shall be deprived of his life or personal liberty except according to procedure established by law" (Constitution of India, 1950: 109). In 2010, the Supreme Court of India "expressed anguish at courts' apathy to the plight of workers being retrenched in the guise of globalization and economic liberalization" (Corbridge et al., 2013). The Supreme Court of India's judgment pointed out:

> It needs no emphasis that if a man is deprived of his livelihood, he is deprived of all his fundamental and constitutional rights and for him the goal of social and economic justice, equality of status or opportunity, the freedoms enshrined in the Constitution remain illusory.
>
> (ibid, 2013)

The idea behind the judgment was to protect the rights of the weaker sections given the clear constitutional mandate. Social justice was a principle mentioned in the Constitution of India. Therefore, it was imperative for the courts to protect the rural poor in cases when the Indian state failed to deliver on a local level. At times, the Supreme Court has intervened and pulled up the federal and provincial governments to act in cases where rural employment policy is not followed in letter and spirit. Thus, India is one of the most vibrant democracies in the world.

MGNREGS is legislation that can be called a social welfare scheme for redistributing the wealth from the rich to the poor. However, social entitlement is not a charity granted by the Indian state. Rather, the entitlements are part of the provisions mentioned in parts III (Fundamental Rights) and IV (Directive Principles of State Policy) of the Indian Constitution that deal with justifiable and unjustifiable rights. The emergence of MGNREGS came as a result of judicial activism that prompted realization of basic rights for the Indian citizens. This would help to transform constitutional promise into a legal entitlement (Pankaj, 2012).

There were contradictions that existed in MGNREGS implementation, which has been highlighted through secondary literature since enactment of legislation in 2009. This chapter is focused on unearthing bottlenecks through primary data obtained by fieldwork in Jharkhand.[2] Why was Jharkhand chosen a site for conducting fieldwork? The mineral-rich province has low rank in the Human Development Index and is considered to be one of the poorest provinces[3] in India. A study by Bhandari and Chakraborty pointed at human development distress in Jharkhand: "at an aggregate level, 50 percent of the districts are identified with poverty levels greater than 40 percent, the inaccessibility of the state makes identification of the poor even more difficult" (Bhandari and Chakraborty, 2014: 1). However, the province continues to grapple with political instability at the provincial level as well as with security threats from Maoist insurgents, who have used the tool of inept governance to expand their cadres (Sareen, 2016).

Meanwhile, the reason to examine local bureaucracy,[4] which constitutes a significant portion of street-level bureaucracy (SLBs), is that it shall provide epistemological and ontological justification to understand the workings of the Indian state at the micro level. Also, there is minimal academic research conducted on understanding the instrumentality of the local bureaucracy in Jharkhand.

MGNREGS: a background

The salient features of MGNREGS are aptly summated by scholars. Each rural household is entitled to a free job card with the photographs of all adult members living in the household. The adult members can be provided employment only if they possess job cards. The government is obliged to provide work within 15 days of sanctioning of a work site, failing which the applicant is entitled to an unemployment allowance. Furthermore, work must be given within 5 kilometers of the applicant's residence, or a 10 percent premium is added to the scheme's wage (Ravallion et al., 2014).

An all-India uniform wage of Rs. 100 per day (USD 5.70 in approximate value) was established in the scheme but was adjusted as per the state-specific inflation. In 2009, the central government uncoupled MGNREGS wages from the state-level statutory minimum wage rates. Based on a schedule of rates, the payment depended on the amount of work done by a person. The act requires the schedule of rates to be set such that an able-bodied worker who works for nine hours, with one hour of rest, can earn the established program minimum wage. In the case of Jharkhand, it was increased over the years from Rs. 100 to Rs. 158 (2014) as per the consumer price index for agricultural laborers (CPIAL) in a particular province. Experts at the National Institute of Rural Development are of the view that provinces were arbitrary in increasing MGNREGS wages. For instance, some provinces increased wages, while others were slow in revision. The beneficiaries wrote a letter to the PM to protest meager increase of Rs. 5 in their wages (Indian Express, 2015). No action was taken by the local authorities on this matter. The beneficiaries were asked to wait for a reply from the prime minister's office, which never came. When MNGREGS started its operation, salaries of the workers were paid in post offices; after reports of corruption came to light, wages were paid directly into their bank accounts.

There were no first aid facilities which were provided at the work site; basic facilities included temporary shades, drinking water packets and childcare assistance for children under age 6.

Several provisions of the act encouraged the participation of women. First, the act mandates one-third of the workers to be women. Second, it ensures equal wages for men and women, with gender-specific productivity norms in the schedule of rates. Finally, the act mandates provision of work locally and childcare facilities. The legislation concentrates on labor-intensive jobs focusing on rural development. Major work carried out under MGNREGS in Jharkhand and other provinces are on water and irrigation activities as well as enhancing connectivity through building roads in the respective village.

Local village governments such as PRIs and the community, play a central role in the implementation. PRI leaders and communities are meant to identify the list of work through discussions in village meetings (*gram sabha*). PRIs also participate in the execution (at least 50 percent work of the is implemented through the *gram panchayat*), supervision and monitoring of work (through social audits).

Ideally, the implementation of the scheme should take place with support from local self-institutions such as PRIs.[5] The PRI leaders and communities are meant to identify the list of work through discussions in the village meetings (*gram sabhas*). PRIs also participated in the execution (at least 50 percent of the work is implemented through the *gram panchayat*), supervision and monitoring of work (through social audits).

A dedicated administrative structure is required for MGNREGS implementation. The act makes specific provision for administrative costs to be borne by the federal government and supplemented by the provincial governments. There is a provision for a dedicated cadre of MGNREGS officials and functionaries at the district, block and *gram panchayat* levels that handle program implementation.

Central and state financing: the central government shares 90 percent of the total expenditure of the program. This includes 100 percent of wage costs and 75 percent of the nonwage component (including materials and wage payments to skilled and semi-skilled workers and most administrative tasks, subject to a maximum limit), working under an assumed 60:40 labor-to-capital ratio. The provinces need to pay for 100 percent of the unemployment allowance costs. The act states that it is the responsibility of the provincial government to pay 100 percent of unemployment allowance costs.

A study on MGNREGS supported enactment of this social welfare legislation, which went a long way in addressing causes of rural distress in India. "The guarantee of work provide valuable insurance against the many risks faced by India's rural poor in their daily lives and help underpin otherwise risky investments" (Ravallion et al., 2014: 6). Scholars argued in defense of the legislation have pointed out at the problems faced by beneficiaries in daily interaction with the Indian state. These include, improving access of information for poor people, which is expected to create adequate demand for work and address supply side factors, for instance, improving administrative capabilities of the local and provincial government (Ravallion et al., 2014).

Dreze's (2011) academic paper "Employment Guarantee and the Right to Work" presented the summarized findings of the sample survey conducted in six states of the Hindi-speaking region (Bihar, Chhattisgarh, Jharkhand, Madhya Pradesh, Rajasthan, and Uttar Pradesh) and concluded that despite some drawbacks, the scheme proved to be beneficial for the poorest of the poor people in these provinces. He noted that 73 percent of the total workforce consisted of individuals belonging to the scheduled castes and tribes. Women had a sizeable presence in the labor force. MGNREGS created useful and sustainable assets that can generate a multiplier effect leading to increased employment opportunities. For example, repairing old water canals and water tanks, afforestation, soil conservation and building roads, directly and indirectly, affect the livelihoods of the rural poor.

Additionally, the MGNREGS wages produced direct impacts, for instance reducing rural migration, realigning the power equations between the landholders and the landless, empowering women, and freeing agricultural laborers from the dependency entrenched in casteist and patriarchal structures.[6] The scheme, most importantly, provides income transfers to poor households during critical times and therefore enables survival, particularly during slack agricultural seasons or years. The book *Politics and the Right to Work: India's National Rural Employment Guarantee Act* presents arguments pertaining to an evaluation of the MGNREGA's performance and the broader debates in Indian politics and the political economy of the Indian state (Jenkins and Manor, 2017). The authors contend that MGNREGS has made a positive impact on a significant section of the rural population; this is by providing employment to over 50 million households annually (Aggarwal, 2019). However, Jenkins and Manor (2017) don't address the issues of economic gains accrued from the infrastructure created under MGNREGS.

Implementation issues

In the first five years of implementation (2006–2010),

> all Indian state and the provincial government spent the funds on the scheme. After that, the excitement slowed down. For Jharkhand, the expenditure was Rs. 57.79 billion (889,199 USD of approximate value) that accounted for 4.67 percent of the India's rural poor; however, the MGNREGS expenditure was 4.48 percent.
>
> (Breman and Varinder, 2012)

Furthermore, the indicators, such as employment generation, wage receipts and impact on poverty, were used to examine the performance of the scheme from 2006 to 2011. Job creation in Jharkhand was pegged at 44.3 man-days per household. Several districts in Jharkhand generated an average or low-level employment (Breman and Varinder, 2012). In India, MGNREGS played an instrumental role in the economic empowerment of women, social inclusion, asset creation, protection against distress migration and increased reserve price of labor (Breman and Varinder, 2012).

However, the bottlenecks in the scheme implementation were highlighted by a study conducted in Jharkhand (Bhatia and Dreze, 2006) titled,

"Employment Guarantee on the ground: Insights from Jharkhand." The study revealed a low awareness among the MGNREGS workers. For example, the job card was a card that individuals or households knew they were supposed to receive, but several beneficiaries were not aware of their entitlements associated with the job card. Most people were mindful of the fact that possessing a job card was a prerequisite to applying for or gaining employment under the legislation.

In Jharkhand, the PRIs were not well established. After the province had been carved out in 2000, no local elections were held for the *panchayats*. In the absence of local institutions, the MGNREGS implementation was difficult because the act mainly required the *gram panchayat* to register and issue job cards. Due to unavailability of the local institutions, implementation was left in the hands of local officials (SLBs).

Bhatia and Dreze (2006) noted a low demand for work because of a lack of understanding of the requirement of filing an application to seek employment. The other reported issues were irregularities and delay in payment of wages, fudging of muster rolls, flawed work measurement, and nonpayment of minimum wages.

The fudging of muster rolls negated the guidelines of MGNREGS and thus did not propagate transparency. The two types of muster rolls were available that were fudged were (a) temporary rolls (notebooks in which the daily attendance of the MGNREGS beneficiaries was maintained) and (b) permanent rolls (notebooks in which the attendance of people who participated in the employment scheme was maintained along with their signatures). Muster rolls were created because most poor beneficiaries did not know how to write or sign. The loophole in this process was that it provided minimal bargaining power to the recipients if their details were entered wrongly. Therefore, the supervisors (local officials) were capable of corrupting the rolls and adding fake names in the permanent muster rolls. This introduced corruption in the system and led to a delay in delivery of wages to the beneficiaries (Breman and Varinder, 2012).

Also, studies have demonstrated the role of fake *gram sabha* in villages that have not played an active role in the planning and implementation in Jharkhand. Middlemen These fictitious *gram sabha* promoted the vested interests of middlemen (local officials). Furthermore, the *bicholias* (middlemen), who act as a conduit between government officials (state) and the MGNREGS beneficiaries, provided the officials their "cut" and earn "commissions" (Bhatia and Dreze, 2006).

Interestingly, the Sareen (2016) study on state interventions[7] (MGNREGS) for the development of the marginalized Ho tribe in Western Singhbhum district in Jharkhand points out district officials instituted electronic payments (MGNREGS) into an individual beneficiary account that did not prevent delay in delivery of wage payments. The sole reason is that the project contractor[8] "exploit some villagers lack awareness of banking system and appropriate profit in their name" (Sareen, 2016: 10). The common criticism that works in MGNREGS amounts to little more than "digging holes and filling them up." Survey in the recent past have shown that 90 percent of respondents considered the public works to be useful or at somewhat useful (Kulkarni et al., 2015). The survey in Jharkhand of almost 1,000 open wells estimated the real rate of return of completed wells being used for irrigation to be close to 6 percent, a respectable investment for any economic investment (Bhaskar et al., 2016).

Methodology and fieldwork challenges

It is essential to highlight the grievances of the informants from implementation so that the identified bottlenecks help answer research questions and provide an understanding on the operationalization of the Indian state at the local level. To translate the narratives in simplistic terms, "qualitative research is conducted through participation/observation method, which is ubiquitously applied in ethnographic studies, followed by qualitative interviewing, focus groups, and language-based approaches to discourse and conversation analysis" (Bryman, 2001: 383). To contextualize the study and look at words rather than data, I employed focus group discussions as a method for data collection. It is a flexible method that involves less time on the field and enables ascertaining substantial information in a short span. Similarly, the focused group discussions form an important part of data collection. A moderator (mate)[9] led ten informant beneficiaries in a village. In the initial conversations, I observed introverted behavior from the informants; however, with time they began to speak up. Due to the paucity of time, I was humbled to respect the informants and their cultural traits and tried not to be intrusive.

Then, the challenge in a focused group discussions is how to manage the group environment? The group environment draws out a variety of perspectives and discussions while providing greater details and uncovering various facets of the issues; an advantage of this research tool is that it generates a larger gamut of perspectives than in-depth interviews with similar informants (Hennink et al., 2011). During the focused group discussions, at times, managing the group, dynamics became a problem. The women informants were more vocal to express their resentment against the role of SLBs and challenges associated with MGNREGS implementation. On the other side, the men informant beneficiaries did not speak openly against the local institutions and its officials.

In Dumargarhi village, the male informants appeared to be reserved during interactions compared to their female counterparts, who vociferously pointed out the functional deficiencies of the Indian state. At that moment, the gatekeeper told me: "when the female join the discussion then males will take a back seat." This aspect was visible in the focused group discussion in the remaining three villages of Jharkhand. After we had completed the discussion with the group, I asked the audience if they had any questions for me. In addition, they were given adequate time to ask questions. Some did raise important issues or sought my clarification on my position as a researcher.

To ensure the reliability of the information gathered in Dumargarhi village, the next day, I attended the conference to set up social audit directorate in Ranchi city. In the conference, the MGNREGS commissioner (a federal bureaucrat) chaired the session to seek feedback from the NGO activists who had assembled from all the districts in Jharkhand. They were divided into small groups and had to submit recommendations to the federal government. I was an observer, and this provided me with a macro view of the problems of scheme implementation in Jharkhand province.

During the fieldwork, four issues came across that may impact the quality of data collected. One was language, the vagaries of transcription, insider and outsider dichotomies and gender positionality. For ethical reasons and to maintain the confidentiality of the informants, I anonymized their names, except four informants who had agreed to quote their real names during the fieldwork. The field notes were transcribed from Hindi to English where modifications occurred.

The focused group discussion brought out these bottlenecks:

1 Delay in receiving wage payment that leads to discontentment toward state and its institutions, especially SLBs.
2 Corruption in the procurement of material cost in MGNREGS (the role of SLBs).
3 Guidelines of the MGNREGS legislation were contravened.

Who are the street-level bureaucrats (SLBs)?

Michael Lipsky defines street-level bureaucrats as "public service workers who interact directly with citizens in the course of their jobs, and who have substantial discretion in the execution of their work" (Lipsky, 2010: 3). SLBs may be teachers, police officers, social workers, judges, civil lawyers, health workers and other public employees. Thus, they can be classified as BDO, BPO, and *rozgar sevak* in the case of MGNREGS. Therefore, these local officials are important conduits between the state (federal and provincial) and the beneficiary. The SLBs are paid from the MGNREGS funds. Their salary is received from the provincial state government (of Jharkhand) through the district office to the *tehsil* (block office).

However, Lipsky maintains that the SLBs face dilemmas from two quarters: one from the public service recipients (beneficiaries) and the second from the civil society activists to prove their efficacy and efficiency of public service. That may be true in MGNREGS implementation. During the fieldwork (Jharkhand), it was found that the beneficiaries and civil society activists created pressure on the SLBs to act as per the regulations mentioned in the legislation.

On the other hand, Lipsky argued in defense of SLBs and their active role in policy implementation. He cited instances from cities in the United States where the role of SLBs was important in facilitating governance at the local level. In the context of MGNREGS implementation in Jharkhand, they acted as conduits between both the federal and provincial states and the MGNREGS beneficiaries. That is where the role of SLBs was perverted.

Interestingly, Lipsky's argument centers on the discretionary power possessed by the SLBs to implement social welfare policies. He called SLBs the real policy makers. He continued to vouch for SLBs and the difficulties they experience in everyday activities to implement policies as per the needs of the subjects. To address the concerns of the citizens, they are "forced to use their discretion and relative autonomy from agency authority (federal state) to ration services to bring the level of services in line with resources" (Smith, 2003 cited in Peter and Pierre, 2003: 356). That is why Lipsky called them the real policy makers. Therefore, in their day-to-day functioning, Lipsky justified the use of discretion and relative autonomy as an attribute for professional training.

In defense of the SLBs, Lipsky argued:

> The essence of street level bureaucracy is that they require people to make decisions about other people. Street level bureaucrats have discretion because the nature of service provision calls for human judgment that cannot be programmed and for which machines cannot substitute.
>
> (Lipsky, 2010: 161)

Lipsky defended SLBs to make a point that his theory was closer to the top-down approach, where the implementation power lay in the hands of street-level bureaucrats.[10] In defense for the SLBs, the aim was to delink policy implementation from top-level functionaries to the lower level bureaucrats so that access to public services to the citizens is improved (Lipsky, 2010). Furthermore, he argued:

> the clients who receive intended benefits obtain the services that are not available elsewhere. So the government agencies have a monopoly on the service, clients may not be able to afford private services, or may not have ready access to them."
>
> (Lipsky, 2010: 54)

Some of the essential services classified under the domain of SLBs are health care and legal services. However, MGNREGS is a rural employment guarantee program that provides 100 days of employment to individuals and/or households in rural India. Similarly, the beneficiaries are paid wages to carry out developmental or public work tasks in their respective villages. So the context of a comparison between Lipsky's SLB and the BDO in MGNREGS is different.

Interestingly, the narrative on SLBs was extended by comparative public administration scholars Meyers and Vorsanger (2003), who believed that the lower-level bureaucrats were important as they created a normative order in society. The contemporary scholarly work on street-level bureaucracy describes them as frontline workers who are the policy makers and integrates them with a new public management approach that is focused on performance and market competition in public services in the Western countries. Bovens and Zouridis (2002) provided a transformed view of discretion and argued,

> Public servants can no longer freely take to streets, but are connected with the organization via the computer. Client data must be filled in, with the help of fixed templates, in electronic forms. The knowledge management systems and digital decision trees have strongly reduced the scope of administrative discretion.
>
> (Bovens and Zouridis, 2002: 177)

In a sense, Lipsky's SLB can be interpreted in Max Weber's understanding as an "administrative unit with the ability to frame rules for the society which for instance guarantee rights and protects its subjects to achieve a good economic life" (King and Le Gales, 2012: 116).

Meanwhile, in India SLBs have not reached a stage where they act in an impartial manner or where their performance can be measured in scientific terms. There remain contradictions in their functioning that ultimately undermine the directives of the federal bureaucracy (Brehm and Gates, 1997). There are a plethora of justifications provided in the scholarly work of Meyers and Vorsanger that justifies the importance of SLBs as leaders who are accountable and/or active and act as an interface between citizens and the state. Post-Lipsky arguments do justify the implementation of public and or social welfare policies through the local officials. Lipsky had understood the role of SLBs from a developed country perspective. Therefore, to contextualize Lipsky's SLB discretion model in India may not be possible, as indicated by fieldwork in Jharkhand.

Meyers and Vorsanger (2003) acknowledged the contradictions in studying the concept of street-level bureaucracy from a positive lens; we need to understand these contradictions rather than to shower accolades on the functioning of street-level red tape. The scholarly discourse on SLBs post-Lipsky painted a grim picture of them. The SLBs were cited as frustrated and powerless cogs in bureaucratic machines, as self-interested bureaucrats whose coping mechanisms frustrate and distort the policy intentions of elected officials.

On the plight of SLBs in performing routines and daily activities, Lipsky argued:

> street level bureaucrats spend their work lives in a corrupted world of service. They believe in themselves to be doing what they can under adverse circumstances and develop techniques to salvage service and decision-making values within the limits imposed upon them by the structure of work. They develop conceptions of their work and of their clients that narrow the gap between their personal and work limitations and service ideal.
>
> (Lipsky, 2010: 12)

Through this, Lipsky justified the presence of the state and its local officers, who even performed in adverse circumstances and were able to implement public policy in letter and spirit.

"Academic studies depict SLBs as heroic local leaders who translate impersonal policy directives for the benefits of their clients" (Meyers and Vorsanger, 2003: 250). The secondary academic literature does not discount frugal theory and methods for studying street-level workers and the failure to contextualize their performance. In the context of MGNREGS, when the regulations of the act are contravened blatantly in Jharkhand, it becomes necessary to look at the academic discourse supporting the empirical data obtained from the fieldwork in Jharkhand.

Tummers and Bekkers (2014) defined Lipsky's argument on discretion provided to the SLBs with the help of a diagram.

Figure11.1a can be inverted as per the negative outcome in MGNREGS implementation. Ironically, discretion can be inverted as presented in Figure 11.1b (Part 2) based on empirical evidence from Jharkhand.

Inversion of discretion

The Lipsky model of discretion on street-level bureaucracy is inverted primarily due to corruption and its intrinsic connection with street level bureaucrats in Jharkhand. The SLB model is inverted based on the testimonies received during the fieldwork in Jharkhand. As mentioned earlier, the scholars held a

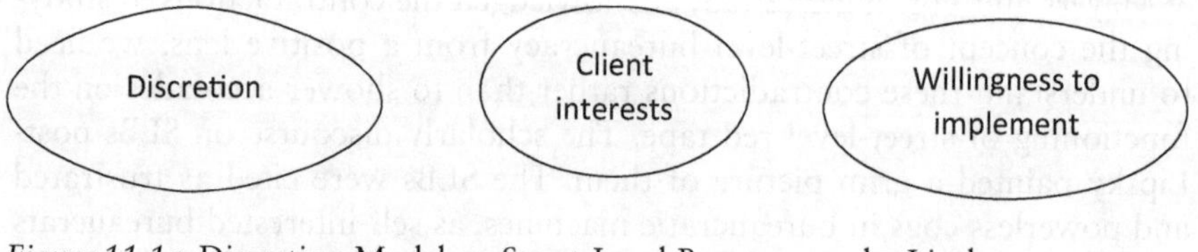

Figure 11.1a Discretion Model on Street-Level Bureaucracy by Lipsky

Source: Prepared by author

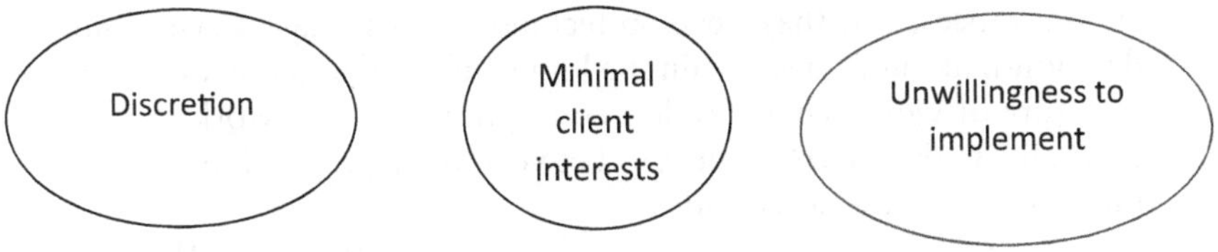

Figure 11.1b Discretion Model on Street-Level Bureaucracy by Lipsky
Source: Prepared by author

positive view of the street-level bureaucrats that acknowledges the genuine hardships through which they implement the social employment welfare scheme on a daily basis. However, the clients/beneficiaries saw SLBs had minimal interest in them and SLBs' unwillingness to apply the scheme in letter and spirit. Lipsky stated: "Clients see unfairness; street-level bureaucrats see rational responses to bureaucratic necessities" (Lipsky, 2010: 116). This argument can be contextualized in the case of Jharkhand, where the local state agencies view rational responses to bureaucratic orders of their superior, and, on the other hand, beneficiaries see the unfairness of the local officials[11] in the day-to-day activities during implementation. SLB informants saw problems in execution and the difficulties (fiscal constraints,[12] personal issues) in which they need to provide access for the beneficiary informants.

The discretion, as depicted in Figure-11.1a, was used by the SLBs to further their vested interests. What were these vested interests? An attempt to answer that is provided in the patronage section of this chapter. Scholars contended that vested interests can influence the program implementation in a negative way and ultimately undermine the effectiveness and democratic legitimacy of the program (Tummers and Bekkers, 2014: 530). However, scholars agree that to follow standards and regulations, street-level bureaucrats need to prioritize regulating the rules and the way they are operated (Maynard-Moody and Portillo, 2010 cited in Tummers and Bekkers, 2014). Incidentally, as per the observations from fieldwork in Jharkhand, the SLBs do not follow the rules and regulations enshrined in the legislation.

Linkage of conceptual approach with empirical observations

Client interests – unwillingness

Lipsky argued:

> SLBs often spend their work lives in a corrupted world of service. They believe themselves to be doing the best they can under adverse

> circumstances, and they develop techniques to salvage service and decision-making values within the limits imposed upon them by the structure of work. They develop conceptions of their work and of their clients that narrow the gap between their personal and work limitations and the service ideal.
>
> (Lipsky, 2010: 12)

In simplistic terms, they work in client interests and often possess lots of freedom and autonomy to implement public policy programs.

Nevertheless, an informant testimony justified the flaws in the Lipsky SLB model. He said: "the street-level bureaucrats handled the MGNREGS implementation in Jharkhand. According to him, these SLBs were the block-level officials, such as the panchayat; rozgar sevak and the mate were involved in what he called the holy nexus of corruption" (Interview correspondence, December 2014).

He stressed that the people lacked trust about their entitlements because of delayed payments. The real face of the Sarkar (government included the mates; *rozgar sevaks*; SLBs) did not perform their job as mandated by the regulations in the Act. These officials were the last point in the delivery mechanism, but they failed to meet the regulations. How does this work? If workers under the scheme required work, then the demand for work was created and the mate was an essential person. He is instrumental in obtaining the entitlements for the villagers in light of their literacy levels being abysmally low. In a real sense, the SLBs were instrumental in hijacking the system to their advantage and did not care about the clients' interests.

Further, a key informant in the study suggested there was operational ineffectiveness in the implementation of the scheme. He clarified that the scheme was relatively autonomous from the top (federal governmental level). The nuts and bolts responsible for MGNREGS implementation were the SLBs, who did not want the program to be a success.

The realist face of SLBs in Jharkhand was contrary to the heroes he made of them. Lipsky was of the view that SLBs worked under adverse circumstances and tried to make the best out of the situations. Conversely, Ramu opined, people lacked trust in the governmental processes due to delay in wage payments and delay in procurement of construction material wells. Also, for them, the face of the "Sarkar" (government) was the mates and *rozgar sevaks*, as they were the last point person in the delivery mechanism. When the workers involved in the scheme required work, the demand was created by the mate,[13] who was a famous person in this scheme and was instrumental in creating demand for work and obtaining the entitlements for the villagers, whose literacy levels are abysmally low. The dependence of beneficiaries on the mate gave him a chance to siphon off money from post office accounts, act as a conduit between SLBs and beneficiaries bungle the attendance sheets and muster rolls and, in the end, exercise patronage in the village. It was observed that in certain cases, the mate acted as an SLB to the beneficiary informant.

Unwillingness to implement

The key feature of the SLB was a desire to implement the public policy.

> At best, street-level bureaucrats invent modes of mass processing that more or less permit them to deal with the public fairly, appropriately, and thoughtfully. At worst, they give into favoritism, stereotyping, convenience, and routinizing – all of which serve their own or agency purposes.
>
> (Lipsky, 2010: 16)

Lipsky in his later writings admitted two ways of understanding SLBs. In the first category came those who engaged with the citizens (social workers and police officer). In the second category, the SLBs interact with the citizens on a daily basis. In that case, they are unable to match the ideal conceptions in practice due to limitations in the work structure. The second conception is relevant for Jharkhand. In a nutshell, there was a willingness to implement, but the work structure does not let them perform their duties in practice. Post-Lipsky the academic discourse paints a contradictory picture of the functioning of SLBs. The lacuna is located in the theory because it fails to address the dimension of SLB behavior across diverse settings (Meyers and Vorsanger, cited in Peters and Pierre, 2003).

Interestingly, the focus group discussion with informants in the Dumargarhi village (Jharkhand) pointed out an inherent discontentment with the Indian state and its local institutions. The reasons were: delay in wage payment and procurement of the material costs. It can be implied there was disconnect between the beneficiary informants of MGNREGS and the SLBs. The latter did not bother to inquire about the progress of work at the work site. During my visit to the Angara block development office, the SLBs pinned the blame on the mate or *rozgar sevak* in the village for faulty implementation. Who are the *rozgar sevak*? They were responsible in the village to ensure timely maintenance of attendance, wage payments, procurement of construction material and supervision of the work sites. Instead, the SLBs began to narrate their grievances (nonpayment of salaries; non-release of funds from the federal government). All this seemed a vicious circle of maintaining government records, shifting blame for non-implementation to either superiors or juniors. In all this, a visible behavior was their unwillingness to implement the program. There was also a nonchalant attitude of the local officers that created discontentment toward the state and its institutions.

It is important to present an extract from the fieldwork as an evidence of the situation at the local level in Jharkhand. For instance, the informants in Dumargarhi village said that work for construction of a well was stalled for few months, and that led to a financial loss for Tinu. The well had to be constructed on his land, but the delay in procurement of the material led him to borrow money from a money lender. That brought him in debt.

Additionally, Dheeraj, who was a locally elected official in the village *panchayat* and an informant, contended:

> There is a 21-day delay in receiving wage payments. From the last six months, we filled the muster rolls, yet did not receive money from the post office. The postal clerk was possibly involved in embezzlement of the funds. In one case, money was withdrawn twice from same person's account. It happened when the person withdrew money from the post office. Instead of one entry in the passbook, two entries were made, and the person only realized it when he came back home. Therefore, despite withdrawing money once, the money was debited twice. We complained but received no response from the post office.

While Dheeraj was interacting, immediately Sanjay started to speak in the focused group discussion. Sanjay reiterated:

> about the delay in wage payments and complained that INR 2500 was withdrawn from his post office account when someone had signed his name. He was not aware how that was possible. If we get MGNREGS work and payment, then it is good. If we do not get it what will we eat and what will we give our children? We protest to the officials, but they give us false assurances we shall get the money soon.

When the things went out of control, he took the help of NGO activists who gathered people with similar issues and organized a protest against the local block office in the Angara. The SLBs pacified them and gave false assurances to address their grievances. The testimony is a justification for the inversion of Lipsky's SLB discretion model and points out in practice how discretion is used as a tool to subvert client interests and promote their vested interests.

Ironically, in the year 2015–2016, there were a whopping INR 105.8 million crore wage payment delays across India. "The monetary fund squeeze sets off vicious cycle-delayed payments leading to lesser demand and still fewer payments resulting in helpless and distress migration" (Dey and Roy, 2016: 1). So the client interests are subverted due to delayed delivery of wage payment. A recent news report on MGNREGA pointed at inability of villagers to withdraw their wages in December 2018, as their accounts were frozen. This has happened primarily because their bank accounts are not linked to *Aadhaar* (Unique Identification Number) without seeking beneficiaries consent. This has caused a delay in receiving wage payment and or led to wages being deposited in the wrong accounts in Jharkhand (Anwar, 2019). The Indian state still claims that 90 percent of wages are paid within 15 days, a recent study done by social activists suggested, more merely 21 percent of payments in 2016, and 32 percent in the first half of 2017 were made in 15 days. (Grover, 2019)

Corruption narrative

Likewise, the image of SLBs is tarnished in a real sense. That is the rationale for the narrative of corruption explained by renowned political anthropologist Akhil Gupta (2012) that lower-level officials are in the line of fire compared to the federal public servants who maintain a benevolent and generous image. This phenomenon points toward a claim made in scholarly literature about a paradox of corruption, wherein the programs meant for poor people fail to deliver on the promises made. A scholar has observed: "The failure of democracy to give effective voice to substantive demands has locked in a vicious cycle that is eroding the very legitimacy of democratic governance" (Heller, 2000: 496).

The narrative on corruption by Gupta holds true in the fieldwork (Gupta, 1995). For instance, an informant said; "I had to pay my personal expenses to procure the material and still did not get the money from the government." Then, Ramu explained that there was violence of corruption performed in a technical way. This builds pressure on an honest mate to work in the interests of the beneficiaries. In general, on an average, 95 percent of mates have connived with local officials (SLBs) and discarded the concerns of the recipients. The testimony is similar to a study on the construction of wells by (Aggarwal et al., 2012) described, "well owners had to pay bribes at various stages of the construction process. Wage payments for laborers tend to be delayed, making them turn to the good owners for interim payments." The study was carried out in the Purio *gram panchayat* in the Ranchi district. The article refers to criticism by scholars (World Bank, 2011 report) of MGNREGA as a make-work scheme and where many public works are washed away the next monsoon. A narrative was provided recently in Jean Dreze article "The digging holes myth" where he does acknowledge corruption leads to delay in the completion of dig wells in Jharkhand (due to delay in wage payment and or delay in procurement of construction material). Nevertheless, "despite considerable hassle at the construction stage, sometimes even leading them to sell their property almost all well owners were glad that they have built well"[14] (Dreze, 2015).

An informant who was a *panchayat* official elaborated later on the intricacies of bribery in the focused group discussion. The process was complicated to detect corruption. For instance, to procure construction material from a government supplier, a TIN[15] (Tax Information Number) was needed. In principle, in each and every village a TIN should be available. However, in practice, the TIN is hardly freely available; that means to procure the material, beneficiaries need to contact the person who possesses it. Here lies the hidden hand of corruption, wherein: "Only when the TIN is provided. The construction material is supplied." Earlier, a 4 to 5 percent commission[16] was cut to a person who had the TIN. Now it is like this, a 14 percent commission[17] to all those who possess the TIN. *Panchayat sevak* and *rozgar sevak* help in the procurement of TIN; then they have a "setting" (implying corruption) to get the TIN number. Corruption is done technically. If I procure material from the TIN person then, I shall pay the money in cash after cutting a 14 percent commission. It is not written in the act but in practical way this sort of corruption. When you ask, they say I send the money to the bank account, but the TIN person will "get it in cash." The vested interests of the SLBs to short-circuit the system are in the form of corruption.

Relevancy of patronage

Chandra described India as a patronage democracy where the elections are conducted through the universal adult franchise but the state has monopoly over a large number of public goods and services, and the allocation depends a lot on the individual discretion of the officers who oversee the implementation of rural employment policy in rural India (Corbridge et al., 2013). On the other side, anthropologist Piliavsky described "patronage, not in the domain of transactions. Rather a moral idiom through transactions is organized through patronal munificence and the client's service (seva) that is selfless acts" (Piliavsky, 2014: 22). Selflessness is explained by the term *mai-baap* (literally mother and father, but it implies patron in this context). Pilivasky delineates the practical terms on which moral logic operates in the context of political patrons and fixers both of extract money and political support in exchange for access to state resources.

Ramu testimony on patronage:

> the job of mate is a "class monitor" in managing the MGNREGS work in the village. The villagers trust the mate more than the provincial bureaucrats. Also, the mate possesses the de-facto power to withdraw money from the post offices and connive with the post office officials to siphon off the money sanctioned by the government. He agreed there was corruption in the delivery of wages. For this, he accused the SLBs who were disconnected with the rural folklore and rarely visited the worksite to oversee the work; they relied on the inputs of the mate, rozgar sevak, and panchayat sevak.

In simple terms, the mate[18] acted as a fixer between SLBs and the beneficiaries and extracted money on the pretext of selfless service for the villagers. It was found out that SLBs proved to be a conduit between the political classes and at times acted as political patrons to the beneficiary informants. An observation made in the fieldwork does point out that SLBs were looked upon as *mai-baap* (patrons) and on whom the rural poor (clients) trusted. That is why the patronage approach that treats it as a moral idiom is relevant in the context of Jharkhand.

Ironically, the implementation of the program was overseen by the SLBs, who continued to control the access to essential inputs for agriculture, such as water, public sector credit, loans, rations of essential commodities (through public distribution system), and implementation of public works (MGNREGS); the allocation of these resources was subject to political discretion (Corbridge et al., 2013).

A keen informant agreed that the role of SLBs was instrumental in creating bottlenecks in full-scale rural employment policy implementation. Thus, an informant revelation provides an answer to the research question that SLBs were involved in corrupt practices. Further, an informant described the creation of Jharkhand province from Bihar in 2000. In that process, the district bureaucratic had a "colonial mind-set and treated people merely as their subjects and considered themselves as lords or rulers." In a real sense, there was the presence of patronage politics at the block level, where the SLBs acted as patrons for the MGNREGS beneficiaries. Informant posited that patronage did exist at the local level, and the real power lay in the hands of elected local officials as well as local bureaucrats. Chandra's argument is that patronage politics leads to ethnic fragmentation and regrouping of people by one ethnic group (Chandra, 2009). This line of argument was not found in the fieldwork and was not a focus of the research project. Nevertheless, the traces of patronage being used as a transactional tool in connection of MGNREGS could not be ruled out. One such instance was in the 2014 federal elections, when the Congress party carried full-page advertisements in the visual and print media across the country showcasing the success of the scheme and taking the credit for being a pro-poor political party. These antics were done to garner votes from poor rural voters in India. Unfortunately, the Congress party lost badly in the general elections and could not encash the patronage built up ever since the program was enacted in the province or the country.

Interestingly, an anonymous testimony revealed the true nature of the Indian state at a micro level:

> Structural bureaucracy and organizational framework are the biggest stumbling block with regards to the MGNREGS implementation. You try social audit, but the structures remained the same, so

> it does not work. Earlier, it was block programme officers, and now, we have the BDO. They are not implementing it at all.

This testimony provides an answer on why SLBs are involved in corruption, and there are systemic problems – structural and organizational framework of bureaucracy. The *asli* (real) face of the *sarkar* (government) is short-circuited by the SLBs to their advantage. Added to this, Saran's testimony concretizes Piliavsky narrative, "patronage is a moral idiom," embedded in the Indian state at a local level.

Current status of MGNREGS implementation

A demand-based social welfare program has been converted into an asset-based program. In 2015, Prime Minister Modi announced,

> My political instincts tell me that MGNREGA should not be discontinued, because it is a living memorial to your failures. After so many years in power, all you were able to deliver is for a poor man to dig ditches a few days a month.
>
> (Grover, 2019)

Such statements by the top political class in recent years have created a budget ceiling on funds for MGNREGS. India needs to allocate 1.7 percent of its GDP. However, the current allocation is 0.25 percent of GDP in 2017–2018, in contrast to 0.58 percent in 2010–11 (ibid, 2019). Early in 2019, economists, social activists and farmers groups have written to the prime minister, seeking more funds and Indian state's renewed political commitment for MGNREGS.

Conclusion

This chapter has brought out the bottlenecks associated with MGNREGS implementation in Jharkhand, and the problems faced by the beneficiaries were highlighted. SLBs were found blocking access to the public service delivery for the intended recipients. Therefore, to assess the impact of SLBs on the implementation, this chapter engaged in a theoretical discourse on street-level bureaucracy and the counter-narratives (corruption; paper state; flailing state; patronage). The conceptual approaches were linked with the empirical observations obtained in the fieldwork from Jharkhand. After finding the relevance to approaches, an attempt was made to contextualize it in terms of operationalization of the Indian state at the local level.

However, Lipsky's (1980) scholarly work on SLBs portrayed a positive impact on improving the effectiveness of public services. The examination of Lipsky discretion model is to project the SLBs as a messiah in the complex governmental processes. He argued that "the process of street-level policy-making does not involve, as might be hoped, the advancement of the ideals many bring to personal service work but rather the development practices that enable officials to cope with the pressures they face" (Hill, 2003: 52).

Lipsky's discretionary model is that there are many pressures under which SLBs have to perform, and that is one reason why their ideas are tweaked ultimately to suit client interests. The SLB model was contested in the case of Jharkhand, where their role created problems instead of paving the way for smooth implementation. Is there an element of inconsistency in Lipsky's analysis of SLBs? Perhaps he wants to convey a message that under steady fiscal pressures and corruption, the SLBs can control public services, then try to match to the service ideals. Herein lays the dilemma of SLBs, where on the one hand they have the freedom to make policy and on the other they work in a relatively independent position.

Interestingly, from the empirical findings, it can be said that SLBs played a negative role in legislation implementation. The negative part is embedded in corruption. That, in turn, leads to a delay in the delivery of wage payments and delays in the procurement of construction material. Finally, the immediate impact is regulations of the act are contravened, and there is a sense of a *mai-baap* (patron) feeling among the MGNREGS workers.

Similarly, the empirical data provides a diametrically opposite picture concerning the role of SLBs in the case of rural employment policy in Jharkhand. The Lipsky model does not fit in the context of either Jharkhand or India. The SLB discretion model is an ideal directive that the street-level bureaucrats need to follow while implementing public policy schemes. However, the empirical observations in Jharkhand point out otherwise. Gupta's narrative about corruption provides vital insights to understand the workings of local bureaucracy that includes the role of SLBs in the implementation of rural employment policy in Jharkhand (Gupta, 2015).

Ironically, the scheme that was designed to uplift the rural poor faces glitches due to the rude attitude of the SLBs. Thereby, the impact of the latter has created a feeling of alienation and discontentment from the Indian state as well as showed by bureaucratic ineffectiveness at the local level. Corbridge et al. (2014) contend that the Indian state controlled access to employment in public work scheme, such as MGNREGS, through "street-level bureaucrats" and blamed the political class for lack of allocation of resources. The study validates this aspect.

Notes

1 The Planning Commission was set up by a resolution of the Government of India in March 1950 in pursuance of declared objectives of the government to promote a rapid rise in the standard of living of the people by efficient exploitation of the resources of the country, increasing production and offering opportunities

to all for employment in the service of the community. The Planning Commission was charged with the responsibility of making assessment of all resources of the country, augmenting deficient resources and formulating plans for the most effective and balanced utilization of resources and determining priorities (Planning Commission website). Incidentally, in 2014, the nomenclature of the commission stands changed. It is now called *NITI Aayog* (Hindi name for Planning Commission).

2 The fieldwork was carried out in four villages: Dumargarhi, Orwabera, Tati and Bisa in the Angara block of Ranchi district.

3 Another study has stated that Jharkhand has one of the "lowest rates of irrigation coverage ranging from 2 percent in some districts to 24 percent in some. Agriculture is thus mostly rain-fed. To top it all, several parts of the province have been affected by severe droughts for several years over the past decade" (Bhaskar et al., 2016: 41).

4 The local bureaucracy or SLBs have to be contextualized from a political science perspective. That is (King and Le Gales, 2012), the state is placed as an administrative unit with the ability to frame rules for society and guarantee rights and protects its subjects and strive toward economic development. This is where the role of bureaucracy (federal, provincial and local) comes into being. The local bureaucracy is assumed to function on the Weberian principles of interdependence, differentiation, legitimacy and autonomy in a defined territory.

5 Implied local self-government bodies from top to bottom are as follows: *zilla panchayat* (district council); *panchayat samiti* (block council); *gram sabha* (village assembly).

6 The chapter does not delve the impact of bottlenecks in the scheme on caste and or gender.

7 Sareen discusses the Saranda Action Plan (SAP), which aimed at bringing security and development in the conflict-hit Saranda forest area in Jharkhand. His study cites an instance of MGNREGS implementation and how "development as security enables the mutual constitution of top-down authorization and inequitable resource access, building an undemocratic local-level state" (2016: 1).

8 This study does not delve into the role of private contractors in the delivery mechanism. These contractors are called fixers, as pointed out by Ward Berenschot. This study is solely concerned with assessing the role and impact of SLBs on the scheme.

9 Mate is in charge in the village for collection of muster rolls and oversaw payment delivery and interaction with the state officials in the village.

10 In the top-down approach, the federal government through its centralized bureaucratic structures carried out public policy implementation. There was a trend in the developed countries like the United States and United Kingdom. So the Lipsky approach came as a reaction against the top-down approach, where the onus of implementing public policies was given to the local rather than the federal bureaucrats.

11 The unfairness can be elaborated as per the impolite behavior observed during the fieldwork in Jharkhand.

12 Delay in payment of wages or procurement of construction material.

13 MGNREGS provisions violation. Ideally the demand is created by the workers and not the local officials.

14 Dreze opines there is no reason for despondency, as MGNREGS provides productive value to the workers and is rather a skill development program for the rural poor. The federal government views how to make the program more productive; however, Dreze argues that it is one of the largest skill development programs of the federal government.

15 The Tax Information Number was necessary to procure material so that the material used in the construction of the well could be accounted for in the Management Information System of MGNREGS website. Ironically, all the details of information were available from a *sarkari* perspective, but in the *asli* perspective, there were difficulties in procuring the TIN, leading to either worker seeking help from suppliers who possessed the TIN. In return, these vendors used this as a tool to conduct corrupt practices in the implementation process.
16 The amount runs in millions. As the funds allocated from the federal government for a province is in billions of rupees. That is translated to millions of rupees at a block level.
17 The actual cost is in millions if a 14 percent commission is siphoned off by the local officials in MGNREGS. It acts as a constant source of income for these corrupt officials who are not paid good salaries and in turn resort to corrupt means to sabotage MGNREGS implementation.
18 It was observed during the fieldwork in Dumargarhi and Tati village; Mate was perceived by the beneficiaries as the local official working at the block level. However, the function of the mate is to provide information to the SLBs and carry out other administrative tasks to facilitate smooth functioning of the scheme.

References

Aggarwal, Ankita. 2019. 'The Political Project of the MGNREGA,' *Economic Political Weekly*, Vol. 54, No. 10, pp. 20–23.

Aggarwal, Ankita, Gupta, Aashish and Kumar, Ankit. 2012. 'Evaluation of NREGA Wells in Jharkhand,' *Economic Political Weekly*, Vol. XLVII, No. 35, pp. 24–27.

Anwar, Tarique. 2019, 'Jharkhand: Aadhar Derails MGNREGA, Wages Deposited in Wrong Accounts,' February 27. www.newsclick.in/mgnrega-workers-not-paid-months-firs-be-lodged-against-pm-modi, last accessed on 05 April 2019

Bhandari, Laveesh and Chakraborty, Minakshi. 2014. 'Spatial Poverty in Jharkhand,' *Mint*, September 6. www.livemint.com/Specials/bqVly6xj4usB3DiTibS3DK/Spatial-poverty-in-Jharkhand.html, last accessed on 15 December 2014.

Bhaskar, Anjor, Gupta, Sunil and Yadav, Pankaj. 2016. 'Well, Worth the Effort-Value of MGNREGA Wells in Jharkhand,' *Economic Political Weekly*, Vol. 51, No. 9, pp. 40–48.

Bhatia, Bela and Dreze, Jean. 2006. 'Employment Guarantee on the Ground: Insights from Jharkhand,' *Economic Political Weekly*, Vol. 41, No. 29, pp. 3198–3202.

Bovens, Mark and Zouridis, Stavros. 2002. 'From Street-Level to System-Level Bureaucracies: How Information and Communication Technology Is Transforming Administrative Discretion and Constitutional Control,' *Public Administration Review*, Vol. 62, No. 2, pp. 174–184.

Brehm, John and Scott, Gates. 1997. *Working, Shrinking and Sabotage Bureaucratic Response to a Democratic Public*. Ann Arbor, MI: University of Michigan Press.

Breman, Jan and Varinder, J. 2012. 'Implementation of MGNREGA in India: A National Overview,' *Humanist Institute for Cooperation with Developing Countries*. The Hague. www.hivos.org/sites/default/files/publications/the_long_road_paper_9x.pdf, last accessed on 10 October 2014.

Bryman, Alan. 2001. *Social Research Methods*. London: Oxford University Press.

Chandra, Kanchan. 2009. 'Caste in Our Social Imagination,' *Indian Seminar*, No. 609.

The Constitution of India. 1950. *Fundamental Rights.* http://india.gov.in/sites/upload_files/npi/files/coi_part_full.pdf, last accessed on 21 May 2015.

Corbridge, Stuart, Harriss, John and Jeffrey, Craig (eds.). 2013. *India Today.* Cambridge: Polity Press.

———. 2014. *India: Economy Politics Society.* New Delhi: Oxford University Press.

Dey, Nikhil and Roy, Aruna. 2016. 'Starving MGNREGA,' *The Indian Express*, March 30. https://indianexpress.com/article/opinion/columns/mgnrega-poverty-india-unemployment-farmer-suicide-drought/, last accessed on 04 April 2019.

Dreze, Jean. 2015. 'The Digging Holes Myth,' *The Indian Express.* http://indianexpress.com/article/opinion/columns/the-digging-holes-myth-2/, last accessed on 20 March 2016.

Dreze, Jean and Khera, Reetika. 2011. 'The Battle for Employment Guarantee,' in Reetika Khera (ed.), *The Battle for Employment Guarantee.* New Delhi: Oxford University Press, pp. 43–80.

Grover Samarth. 2019. 'Another Lifeline Smothered: MGNREGA in Shambles,' *Newsclick.in.* www.newsclick.in/another-lifeline-smothered-mgnrega-shambles, last accessed on 5 April 2019.

Gupta, Akhil. 1995. 'Blurred Boundaries: The Discourse of Corruption, the Culture of Politics, and the Imagined State,' *American Ethnologist*, Vol. 22, No. 2, pp. 375–402.

———. 2012. *Red Tape.* London: Duke University Press.

Gupta, Akhil, Nugent, David and Sreenath, Shreyas. 2015. 'State, Corruption, Postcoloniality: A Conversation with Akhil Gupta on the 20th Anniversary of "Blurred Boundaries",' *American Ethnologist*, Vol. 42, No. 4, pp. 581–591.

Heller, Patrick. 2000. 'Degrees of Democracy- Some Comparative Lessons from India,' *World Politics*, Vol. 52, No. 4, pp. 484–519.

Hennink, M., Hutter, I. and Bailey, A. 2011. *Qualitative Research Methods.* Los Angeles: Sage Publications.

Hill Michael, Hupe Peter, 2003, *Implementing Public Policy: Governance in Theory and Practice.* London: Sage Publications.

Jenkins, Rob and Manor, James. 2017. *Politics and the Right to Work: India's National Rural Employment Guarantee Act.* New York: Oxford University Press.

King Desmond, Le Gales Patrick. 2012 'State,' in A. Amenta, K. Nash and A. Scott (eds.), *Companion to Political Sociology.* Oxford: Wiley-Blackwell.

Kulkarni, Ashwani, Ranaware, Krushna, Narayanan, Sudha and Das, Upasak. 2015. 'MGNREGA Works and Their Impacts,' *Economic & Political Weekly*, Vol. 50, No. 13, pp. 53–61.

Lipsky, Michael. *1980. Street-Level Bureaucracy: Dilemmas of the Individual in Public Services.* New York: Russell Sage Foundation.

———. 2010. *Street-Level Bureaucracy.* New York: Russel Sage Publications.

Mahatma Gandhi National Rural Employment Guarantee Scheme. 2015. 'May Day Protest: Why MGNREGA Workers in Jharkhand Are Returning Rs 5 to PM Modi,' *The Indian Express.*

Maynard-Moody, S. and Portillo, S. 2010. 'Street-Level Bureaucracy Theory,' in *Oxford Handbook of American Bureaucracy.* Oxford University Press.

Meyers, M. K. and Vorsanger, S. 2003. 'Street-Level Bureaucrats and the Implementation of Public Policy,' in B. Guy Peters and J. Pierre (eds.), *Handbook of Public Administration.* London: Sage, pp. 245–254.

Pankaj, Ashok. 2012. *Right to Work and Rural India Working of the Mahatma Gandhi National Rural Employment Guarantee Scheme (MGNREGS)*. New Delhi: Sage Publications.

Peters, Guy B. and Pierre, Jon. 2003. *Handbook of Public Administration*. London: Sage Publications.

Piliavsky, Anastasia (ed.). 2014. *Patronage as Politics in South Asia*. Cambridge: Cambridge University Press.

Planning Commission. The Government of India. 2008. 'Eleventh Five Year Plan 2007–12,' *Inclusive Growth*. Vol. 1. New Delhi: Oxford University Press.

Ravallion, Martin, Dutta, Puja, Murgai, Rinku and Walle, Van der Dominique. 2014. *Right to Work? Assessing India's Employment Guarantee Scheme in Bihar*. Washington, DC: The World Bank.

Sampath, G. and Rukmini, S. 2015. 'Is MGNREGA Is Step Up for a Failure?' *The Hindu*, May 31. www.thehindu.com/sunday-anchor/is-the-mgnrega-being-set-up-for failure/article7265266.ece, last accessed on 01 March, 2019.

Sareen, Siddharth. 2016. 'Seeing Development as Security: Constructing Top-Down Authority and Inequitable Access in Jharkhand,' *South Asia Multidisciplinary Academic Journal* [Online], 13. http://samaj.revues.org/4146, last accessed on 22 April 2016.

Smith, R. Steven. 2003. 'Street-Level Bureaucracy and Public Policy,' in Guy B. Peters and Jon Pierre (eds.), *Handbook of Public Administration*. London: Sage Publication.

Social Protection for a Changing India, Volume II. 2011. Washington, DC: World Bank.

Tummers, Lars and Bekkers, Victor. 2014. 'Policy Implementation, Street-level Bureaucracy, and the Importance of Discretion,' *Public Management Review*, Vol. 16, No. 4, pp. 527–547.

INDEX